Numerical Simulation in Fluid Dynamics

Mathematical Modeling and Computation

About the Series

The SIAM series on Mathematical Modeling and Computation draws attention to the wide range of important problems in the physical and life sciences and engineering that are addressed by mathematical modeling and computation; promotes the interdisciplinary culture required to meet these large-scale challenges; and encourages the education of the next generation of applied and computational mathematicians, physical and life scientists, and engineers.

The books cover analytical and computational techniques, describe significant mathematical developments, and introduce modern scientific and engineering applications. The series will publish lecture notes and texts for advanced undergraduate- or graduate-level courses in physical applied mathematics, biomathematics, and mathematical modeling, and volumes of interest to a wide segment of the community of applied mathematicians, computational scientists, and engineers.

Appropriate subject areas for future books in the series include fluids, dynamical systems and chaos, mathematical biology, neuroscience, mathematical physiology, epidemiology, morphogenesis, biomedical engineering, reaction-diffusion in chemistry, nonlinear science, interfacial problems, solidification, combustion, transport theory, solid mechanics, nonlinear vibrations, electromagnetic theory, nonlinear optics, wave propagation, coherent structures, scattering theory, earth science, solid-state physics, and plasma physics.

Jianke Yang, *Nonlinear Waves in Integrable and Nonintegrable Systems*

A. J. Roberts, *Elementary Calculus of Financial Mathematics*

James D. Meiss, *Differential Dynamical Systems*

E. van Groesen and Jaap Molenaar, *Continuum Modeling in the Physical Sciences*

Gerda de Vries, Thomas Hillen, Mark Lewis, Johannes Müller, and Birgitt Schönfisch, *A Course in Mathematical Biology: Quantitative Modeling with Mathematical and Computational Methods*

Ivan Markovsky, Jan C. Willems, Sabine Van Huffel, and Bart De Moor, *Exact and Approximate Modeling of Linear Systems: A Behavioral Approach*

R. M. M. Mattheij, S. W. Rienstra, and J. H. M. ten Thije Boonkkamp, *Partial Differential Equations: Modeling, Analysis, Computation*

Johnny T. Ottesen, Mette S. Olufsen, and Jesper K. Larsen, *Applied Mathematical Models in Human Physiology*

Ingemar Kaj, *Stochastic Modeling in Broadband Communications Systems*

Peter Salamon, Paolo Sibani, and Richard Frost, *Facts, Conjectures, and Improvements for Simulated Annealing*

Lyn C. Thomas, David B. Edelman, and Jonathan N. Crook, *Credit Scoring and Its Applications*

Frank Natterer and Frank Wübbeling, *Mathematical Methods in Image Reconstruction*

Per Christian Hansen, *Rank-Deficient and Discrete Ill-Posed Problems: Numerical Aspects of Linear Inversion*

Michael Griebel, Thomas Dornseifer, and Tilman Neunhoeffer, *Numerical Simulation in Fluid Dynamics: A Practical Introduction*

Khosrow Chadan, David Colton, Lassi Päivärinta, and William Rundell, *An Introduction to Inverse Scattering and Inverse Spectral Problems*

Charles K. Chui, *Wavelets: A Mathematical Tool for Signal Analysis*

Numerical Simulation in Fluid Dynamics

A Practical Introduction

Michael Griebel
Universität Bonn
Bonn, Germany

Thomas Dornseifer
Technische Universität München
München, Germany

Tilman Neunhoeffer
Technische Universität München
München, Germany

siam.

Society for Industrial and Applied Mathematics
Philadelphia

10 9 8 7 6 5 4

Library of Congress Cataloging-in-Publication Data

Griebel, Michael.
Numerical simulation in fluid dynamics : a practical introduction / Michael Griebel, Thomas Dornseifer, Tilman Neunhoeffer.
p. cm. – (SIAM monographs on mathematical modeling and computation)
Includes bibliographical references and index.
ISBN 978-0-898713-98-5 (pbk.)
1. Fluid dynamics–Mathematical models. I. Dornseifer, Thomas. II. Neunhoeffer, Tilman. III. Title. IV. Series.
QA911.G718 1997
532'.05'0113–dc21 97-40624

Contents

Preface

The numerical simulation of physical phenomena requires the observations and models of the natural scientist, the technical expertise of the engineer, the numerical methods of the mathematician, and the modern techniques and computers of the computer scientist. Interdisciplinary cooperation among these scientific fields will be necessary in order to significantly extend our capacity to reproduce and forecast physical processes on the computer. Expensive experiments are increasingly being replaced by computer simulations. Moreover, simulation enables the examination of processes that cannot be experimentally tested. The development of new products is being accelerated by the elimination of these costly physical experiments, and product quality is being enhanced by the ability to investigate previously inaccessible phenomena. In the future, numerical simulation will emerge as a key technology.

As a consequence, mastery of all aspects of numerical simulation will soon become indispensible in research and development; among these aspects we specifically mention *modeling*, *discretization*, the development of *efficient solution methods* and *fast solvers*, *parallelization* of the new algorithms, and *visualization* of the computed results. Training in numerical simulation must therefore receive greater attention at the university level.

Elements of numerical simulation may be conveyed to students in mathematics, computer science, or natural and engineering science programs; much, however, remains to be done to achieve a truly cross-disciplinary and application-oriented approach to training. Specifically, while computer scientists are skilled in handling data structures, parallel algorithms, and techniques for visualizing computed data, they are usually less familiar with the underlying mathematical models and their discretization. Mathematicians, for their part, are proficient in mathematical and numerical analysis, i.e., in investigating whether the equations comprising the mathematical model possess solutions and in devising algorithms that compute approximations to these solutions. They are not, however, instructed in how to efficiently implement their algorithms on high-performance computers or in the interpretation of the resulting data. The latter is one of the strengths of the engineers, who can often draw from their experience with physical experiments when constructing numerical algorithms.

It is against this backdrop that we conducted the project class "Scientific Computing and Visualization" at the Technische Universität München for the first time in the summer semester of 1994, in association with the Bavarian Consortium for High-Performance Scientific Computing (Bayerischer Forschungsverbund für Technisch-Wissenschaftliches Hochleistungsrechnen (FORTWIHR)). This class was intended to familiarize students with the essential steps involved in numerical simulation, using fluid dynamics as an example. This project class forms the basis for this book.

After a brief introduction to mathematical modeling in Chapter 2, the central chapter, Chapter 3, begins with a general description of the discretization of differential equations by finite differences, which is then applied to the *Navier–Stokes equations*, the governing equations of fluid mechanics. This is followed by a simple algorithm for the solution of the resulting discrete system of equations.

To enable the interpretation of the flood of computed data, Chapter 4 provides some common scientific visualization techniques; Chapter 5 displays the results of applying the described solution algorithms to various standard problems from fluid dynamics. After an excursion into the simulation of free boundary value problems (Chapters 6 and 7), we turn to parallelization techniques which enable us to solve larger problems and which at the same time are aimed at accelerating the computations (Chapter 8). In Chapters 9 and 10 we extend our mathematical model to include temperature and turbulence and discretize the new equations. We then offer suggestions on extending the program to three space dimensions in Chapter 11. These suggestions are accompanied by several simulation examples.

This book is written on an advanced undergraduate level and addresses both engineers and applied mathematicians. It is also well suited to practitioners seeking basic insights into computational fluid dynamics. The aim of the book is to present—in compact form—the individual steps involved in the numerical simulation of fluid flow, to enable readers to write their own flow simulation programs, and, by presenting various simulation results, to motivate them to perform their own numerical experiments. To facilitate this, each chapter contains, besides descriptions of models and algorithms, practical advice on their implementation, so that a flow simulation code can be written from scratch. Those not interested in the derivation of the mathematical model may skip Sections 2.2, 9.2, and 9.3.

The code for the numerical solution of two-dimensional flow problems is also available from the ftp site `ftp.lrz-muenchen.de`, where it resides in the directory

`pub/science/fluiddynamics/cfd/NaSt2D`.

We thank Prof. Christoph Zenger of the Bavarian Consortium FORTWIHR for his numerous suggestions and support, Prof. Franz Durst from the FAU Erlangen and Dr. Michael Schäfer from the TH Darmstadt for valuable insights from the engineers' point of view, our colleagues Dr. Hans-Joachim Bungartz, Thomas Schiekofer, and Dr. Stefan Zimmer for many helpful hints, our colleague Dr. Walter Huber for advice on parallelization and turbulence, and our students Michael Bader, Andreas Dehmel, Thomas Gerstner, Stephan Knapek, Florian Meier, and

Erwin Wagner for coding, running various model problems, and generating the figures.

Furthermore, we thank the Bayerische Forschungsstiftung and the Bayerisches Staatsministerium für Unterricht, Kultus, Wissenschaft und Kunst, which, by forming the Bavarian Consortium FORTWIHR, made possible our project class and hence this book.

Finally, we would like to thank Oliver Ernst of the TU Bergakademie Freiberg for the translation of the German edition and for many constructive and useful suggestions.

Michael Griebel, Thomas Dornseifer, and Tilman Neunhoeffer
Bonn and Munich

Notation

Two-dimensional notation is given here. In three dimensions, a third component should be added where appropriate.

$\vec{a} \cdot \vec{b}$	$= \sum_{i=1}^{2} a_i\, b_i$: scalar product of the vectors $\vec{a} = (a_1, a_2)^T$ and $\vec{b} = (b_1, b_2)^T$
$\vec{a} \otimes \vec{b}$	tensor product of two vectors: yields a matrix with entries $(\vec{a} \otimes \vec{b})_{i,j} := a_i\, b_j$
$\vec{n}$	$= (n_x, n_y)^T$: exterior unit normal vector
$\dfrac{\partial f}{\partial x}$	first partial derivative of function f with respect to x
$\dfrac{\partial^2 f}{\partial x^2}$	second partial derivative of function f with respect to x
$\mathrm{grad} f$	$= \left(\dfrac{\partial f}{\partial x}, \dfrac{\partial f}{\partial y}\right)$: gradient of $f : \mathbb{R}^2 \to \mathbb{R}$
$\dfrac{\partial f}{\partial n}$	$= \mathrm{grad} f \cdot \vec{n}$: directional derivative of f in the direction of the exterior unit normal
$\mathrm{div} \vec{u}$	$= \dfrac{\partial u_1}{\partial x} + \dfrac{\partial u_2}{\partial y}$: divergence of $\vec{u} : \mathbb{R}^2 \to \mathbb{R}^2$
Δf	$= \dfrac{\partial^2 f}{\partial x^2} + \dfrac{\partial^2 f}{\partial y^2}$: Laplacian of $f : \mathbb{R}^2 \to \mathbb{R}$
Ω	flow domain, subset of $\mathbb{R}^2$
Γ	$= \partial\Omega$: boundary of domain Ω
$\mathcal{G}$	$= [0, a] \times [0, b]$: rectangular domain
$\mathcal{H}$	obstacle domain

$\vec{x}$	$= (x, y)^T$: position vector
$\vec{u}$	$= (u, v)^T$: velocity vector
u	velocity component in x-direction
v	velocity component in y-direction
p	pressure
ϱ	density
$\vec{g}$	$= (g_x, g_y)^T$: body forces, e.g., gravity
μ	dynamic viscosity; cf. (2.17) on page 17
ν	kinematic viscosity; cf. (2.17) on page 17
Re	$= \dfrac{\varrho_\infty u_\infty L}{\mu}$: Reynolds number; cf. (2.20) on page 19
Fr	$= \dfrac{u_\infty}{\sqrt{L \, \|\vec{g}\|}}$: Froude number; cf. (2.20) on page 19
ψ	stream function; cf. (4.4) on page 60
ζ	vorticity; cf. (4.5) on page 60
t	time
$\boldsymbol{\sigma}$	$= -p\mathbf{I} + \boldsymbol{\tau}$: stress tensor; cf. (2.12) on page 17
$\boldsymbol{\tau}$	viscous part of stress tensor; cf. (2.12) on page 17
$\boldsymbol{\delta}$	strain tensor; cf. (2.13) on page 17
$\mathbf{I}$	unit tensor
$i_{\max}$	number of cells in x-direction
$j_{\max}$	number of cells in y-direction
δx	mesh width in x-direction
δy	mesh width in y-direction
δt	time step size
$p_{i,j}$	discrete value of pressure at cell center $((i - 0.5)\delta x, (j - 0.5)\delta y)$

$u_{i,j}$	discrete value of horizontal velocity at edge midpoint $(i\,\delta x, (j-0.5)\delta y)$
$v_{i,j}$	discrete value of vertical velocity at edge midpoint $((i-0.5)\delta x, j\,\delta y)$
$f^{(n)}$	discrete value of function f at time t_n
F, G	cf. (3.29) on page 33
rhs	right-hand side of pressure Poisson equation
r	residual of pressure Poisson equation; cf. (3.45) on page 37
$\\|r\\|$	residual norm; cf. page 38
eps	value for stopping criterion for SOR iteration
$it_{\max}$	maximal number of SOR steps
γ	upwind discretization parameter
τ	safety factor for stepsize control
ω	SOR parameter
T	temperature
e	internal energy; cf. page 129
h	$= e + \frac{p}{\varrho}$: enthalpy; cf. (9.13) on page 129
β	$= -\frac{1}{\varrho}\frac{\partial \varrho}{\partial T}$: coefficient of thermal expansion; cf. page 129
c_p	$= \left.\frac{\partial h(p,T)}{\partial T}\right\|_{p=\text{const}}$: specific heat at constant pressure; cf. page 129
κ	thermal conductivity; cf. page 130
α	$= \frac{\kappa}{\varrho_\infty c_p}$: thermal diffusivity; cf. (9.17) on page 130
Pr	$= \frac{\nu}{\alpha}$: Prandtl number; cf. (9.5) on page 125
Gr	$= \frac{\|\vec{g}\|\beta\,(T_2 - T_1)L^3}{\nu^2}$: Grashof number; cf. (9.6) on page 126
Ra	$= Pr\; Gr$: Rayleigh number; cf. (9.6) on page 126

Nu	$= \dfrac{Q_{\text{convection}}}{Q_{\text{heat diffusion}}}$: Nusselt number; cf. (9.7a) on page 126
q'''	heat source
$\vec{q}''$	heat flux vector
$\dfrac{D}{Dt}$	$= \dfrac{\partial}{\partial t} + \vec{u} \cdot \text{grad}$: material derivative
$\tilde{F}$, $\tilde{G}$	cf. (9.26) on page 134
H	heat function; cf. (9.29) on page 137
k	turbulent kinetic energy; cf. (10.11) on page 160
ε	rate of dissipation of turbulent kinetic energy; cf. (10.11) on page 160
$\langle . \rangle$	averaging operator
U, V, P, G_x, G_y	averaged quantities u, v, p, g_x, g_y; cf. (10.2) on page 157
R	$= -\langle \vec{u}' \otimes \vec{u}' \rangle$: Reynolds stress tensor; cf. (10.8) on page 158
$\mathcal{R}$	approximate Reynolds stress tensor
ν_T	(turbulent) eddy viscosity; cf. (10.12) on page 160
ν^*	$= \nu + \nu_T$

Chapter 1

Numerical Simulation—a Key Technology of the Future

1.1 Physical Experiments, Mathematical Modeling, and Numerical Simulation

The central task in the natural sciences lies in describing reality as accurately as possible in order to better understand natural phenomena and thus gain insight into the behavior of objects under given conditions. This involves the examination of vastly different phenomena occurring on scales of varying orders of magnitude ranging from the investigation of the nature of matter in quantum mechanics to studies of the origin of the universe. In the engineering sciences, research is conducted with the specific purpose of developing new products and optimizing existing products, e.g., with regard to their performance and energy consumption.

Both branches of science have fundamentally changed our lives in the last few centuries; one need only recall such ground-breaking inventions as the printing press, the steam engine, or electrical power. Today, we can hardly imagine life without newspapers, automobiles, refrigerators, and light bulbs. And, not least of all, the development of computers has revolutionized our society in the preceding decades and will continue to change it in the future.

In the past, there have been two methodical approaches to uncovering the laws of nature: the practical and the theoretical. The *practical approach* seeks to discover physical laws through observation aided by experiments and various devices and measuring instruments. Galileo Galilei, with his falling experiments conducted from the leaning tower of Pisa, during which he is said to have discovered that bodies of different weight fall to the ground with the same velocity, is regarded as the founder of experimental physics and thereby as the first representative of the practical approach.[1]

Today, for example, we examine the influence or noninfluence of gravity on crystal growth in space stations, the effects and side effects of drugs in animal

[1] According to [Hermann, 1980], however, the reports on Galileo's experiments are more of a legend. In particular, it is said that Galileo initially claimed to have often observed a lead ball falling to the ground much faster than a wooden ball. He only later revised this statement and, in thought experiments, arrived at the conclusion that all bodies fall to the ground equally fast in a vacuum.

tests, and the effect of a vehicle's geometry on its drag coefficient in wind tunnels.

The theoretical approach converts the laws of nature to relationships between mathematical quantities, most often employing the language of differential and integral calculus to describe how certain quantities change depending on others. According to the well-known anecdote, a falling apple convinced Sir Isaac Newton that the same force of gravity must govern the entire cosmos; this lead to his development of the theory of gravitation. Newton also described the motion of solid bodies with three laws that now bear his name. James Maxwell is credited with discovering the equations governing electromagnetic fields, and Albert Einstein developed his famous theory of relativity while sitting at his desk. The Navier–Stokes and Euler equations form the basis of the mathematical treatment of fluid flow, which describes the dependence of velocity and pressure on space and time.

Both approaches, however, have their shortcomings. In certain areas, performing physical experiments, such as investigating the effects of an oil spill or an accident in a nuclear reactor, is precluded a priori for reasons of safety. Often, measurements cannot be carried out due to the extremely long or short duration of the experiment or if the quantities involved are too small (electric currents in a microprocessor) or too large (the origin of stars and galaxies). Moreover, many experiments entail a very elaborate setup and permit measurements at only a few points, as is the case, e.g., in weather research.

On the other hand, mathematical equations that describe the physical world with reasonable accuracy are usually so complex that analytical solutions can no longer be obtained. Often an exact solution can be found only for considerably simplified models, such as those resorting to special symmetries or ignoring couplings between certain quantities.

Beside the practical and theoretical approaches, *numerical simulation* has established itself in recent years as a third approach connecting the two traditional ones. Numerical simulation is characterized by the following procedure. From observations of the real world, "theoreticians" derive mathematical equations valid at all (infinitely many) points in space and time. These equations are then discretized, i.e., considered at only a finite number of selected points. At these points, the underlying continuous equations are solved approximately. This implies that physical reality is simulated more accurately as these discretization points are spaced more densely. Recent dramatic improvements of computers in terms of memory size and computing speed permit ever more realistic simulations, so "practitioners" are increasingly able to reproduce their experiments on a computer. Modifications in these experiments can be made with just a few changes in a computer program rather than the often costly and time-consuming changes to an experimental apparatus that were previously necessary. Computed data are then processed by visualization techniques so that it can be interpreted.

Many shortcomings of both the experimental and the theoretical approaches may be overcome by numerical simulation. Access to phenomena which could hitherto not be examined can be made possible, and costly experiments can be avoided. Furthermore, test series may be optimized by quick repetitions with slightly altered parameters or geometries, and more data become available than could ever be collected in traditional experiments. In addition, numerical simulation offers us at least approximate access to the solution of the equations comprising the underlying mathematical models.

Of course, a number of requirements are imposed on each phase of numerical simulation. First, the mathematical model should describe reality as accurately as possible while still remaining solvable. Moreover, the discrete model must approximate the continuous model well. The resulting necessarily high resolution of space by discrete points leads to extremely demanding requirements on computer memory and computation time, in particular when time-dependent processes in three space dimensions are being simulated. Depending on the particular problem under consideration, the solution of the discrete problem requires the execution of many nested loops which may involve time-dependence, nonlinearities, and the solution of large linear systems of equations.

A current topic of research in numerical simulation, for example, is the search for methods which can rapidly solve the discrete equations (multigrid, multilevel, and multiscale methods) and those which can accurately approximate the solution of the continuous problem with minimal memory requirements (adaptive methods, error estimators to determine regions to be locally refined). Parallelization is a further option for solving large problems. It involves the distribution of the problem onto several processors (e.g., by means of domain decomposition techniques), which then concurrently compute parts of the complete solution.

But merely computing the solution is not enough. Interpreting the vast quantities of numbers calls for suitable visualization methods to present the data in meaningful ways. In particular, arranging the results of three-dimensional calculations in an organized fashion is not an entirely easy task.

Ultimately, the computed results must also be compared with those of the corresponding physical experiments and, if necessary, the numerical algorithm must be adjusted (improved discretization via higher resolution or higher order of convergence) or the mathematical model modified by adjusting certain parameters or even by altering some of the equations.

An outline of the individual steps involved in numerical simulation is given in Figure 1.1.

Currently, numerical simulation is employed in many scientific and industrial areas, e.g., mechanical engineering, where the properties of elastic solids are investigated in order to design safe vehicles requiring a minimal amount of material or to analyze and improve the stability of structures. In chemical applications, numerical simulation is used to optimize the reactions of different substances occurring, e.g., in combustion processes. Further applications include the investigation of melting and coating processes, crystal growth, weather prediction, and the optimization of energy consumption by intelligent controlling devices. In nuclear physics, the collision of atomic nuclei and the bonding energy of electrons is computed, while, in astrophysics, scientists simulate the nuclear fusion processes

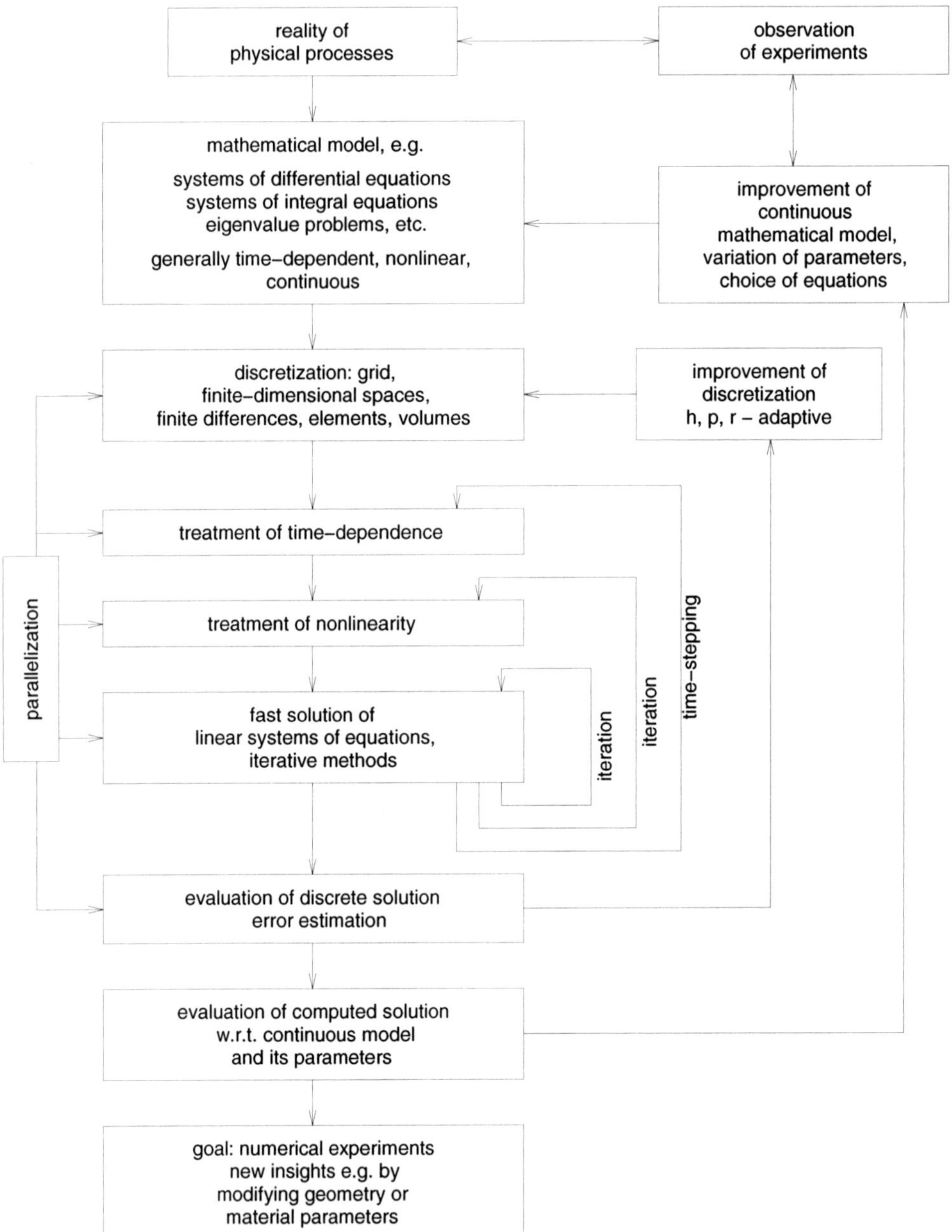

FIG. 1.1. *Typical procedure in numerical simulation.*

taking place in the sun and determine the date when it will be extinguished. A good overview of the manifold applications of numerical simulation can be found in [Kaufmann & Smarr, 1993], and an introduction to scientific computing as well as the discretization of differential equations, in [Golub & Ortega, 1992].

1.2 Fluids and Flows

An important application area for numerical simulation is the investigation of the behavior of fluid flow. We encounter flows every day, often without being aware of it. The behavior of liquids and gases, both of which are considered *fluids*,[2] can be observed in almost all areas of life ranging from complex technical applications to the more mundane situations of daily life.

Everyone has at some time or another admired the interesting patterns forming in their morning coffee cup after adding milk and slowly stirring, or has observed the eddies and waves caused by water flowing in or draining out of a bathtub. The smoke rising from a candle or cigarette and the bubbles ascending in carbonated beverages are further examples.

Outdoors, we can admire a flowing stream, a plummeting waterfall, and, in the sky, the transformation of puffy little white clouds into heavy, anvil-shaped thunderstorm clouds. Driven by the convection within the earth's liquid interior, the continental plates, afloat on the earth's mantle, are constantly drifting apart or colliding to form new mountain belts. Not least of all, the behavior of flows plays a huge role in engineering applications. The drag coefficient of a car, which quantifies the resistance air poses to a moving automobile, is a familiar quantity. Similarly, the construction of modern aircraft would be impossible without detailed knowledge of the flow around the wings.

All these phenomena are caused by various processes taking place within different fluids. The issues to be considered are the interactions between the different fluid particles as well as the forces between moving fluids and solid bodies at rest or between a moving solid body and a fluid at rest, respectively. The source of the occurring forces is a physical property of fluids known as their *viscosity*, which generates frictional forces acting on the fluid that—in the absence of external forces—cause a fluid in motion to eventually come to rest. As an example, we again cite coffee in a cup: after stirring, it doesn't take long for the rotation of the fluid to die down. The coffee has come to rest as the result of internal friction.

To explain these phenomena, we imagine the fluid as consisting of individual layers which can slide over each other, much like a deck of playing cards. These layers are first set in uniform motion. If the forward motion of the bottom layer is suddenly stopped, the layers above continue to slide forward due to their *inertial force*. This force is opposed by friction against the layer below, causing the upper layers to move somewhat further than the lower ones and resulting in roughly

[2] Fluids are substances which, in contrast with solid materials, cannot resist shear stress when at rest.

the situation depicted in Figure 1.2. In this way, the force acting on the bottom layer is transmitted to the other layers through friction. Flows adhering to this idealization are called *laminar* flows,[3] as opposed to *turbulent* flows, in which the particles belonging to different sheets may become mixed due to very small friction, thus leading to an increased effective viscosity (called the turbulent eddy viscosity).

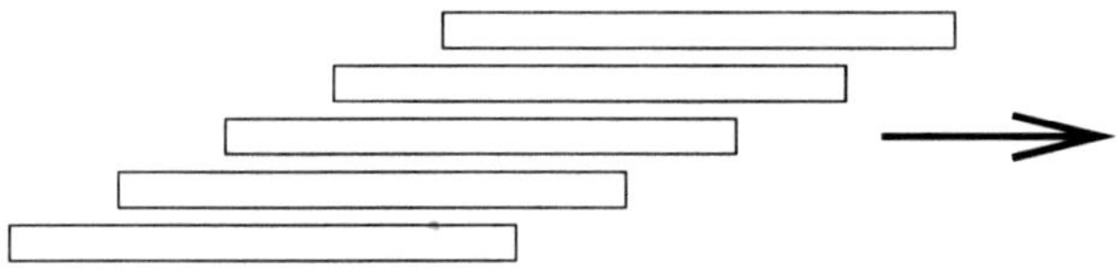

FIG. 1.2. *Laminar flow.*

In highly viscous fluids such as honey, the frictional forces are very strong, so that the different layers come to rest earlier than for less viscous fluids such as water or air. Also, it clearly requires more force to move a spoon through honey than through air. In gases the internal friction is so small that it is often neglected altogether in the model, i.e., in the description of their physical properties by mathematical equations. In their idealized form, gases are regarded as *inviscid* fluids whose behavior is described by the Euler equations in the discipline known as *gas dynamics.*

As already mentioned, the motion of a fluid is determined by the two elementary properties, viscosity and inertia. The relative magnitude of these two properties is measured by a dimensionless parameter named in honor of the British physicist Osborne Reynolds which depends on the velocity of the fluid, its viscosity, and the size of the flow region.

For a long time during the 18th and 19th centuries, scientists in the field of *hydrodynamics*, concerned with the mathematical description of fluid flows, believed that internal friction in water could be neglected. This permitted the explanation of several phenomena such as the generation of waves and the formation of convection cells; in the case of flow past an obstacle, however, the predictions of the hydrodynamicists did not agree with the empirical results of *hydraulics.* The German physicist Ludwig Prandtl resolved this contradiction with his boundary layer theory, in which friction is only considered in a thin layer close to a wall—the so-called boundary layer. In this layer the inertial forces are smaller since the fluid is flowing considerably more slowly than in the interior, coming to rest at the wall itself. The description of flows given by the Navier–Stokes equations accounts for friction throughout the entire flow domain, thus also modeling more viscous fluids. Analytical solutions of these equations, however, can only be obtained under strongly simplifying assumptions.

Let us consider the flow in a river with a narrow section, i.e., an obstacle. If the flow velocity is small, then the viscous forces outweigh the inertial forces, and friction is able to slow the water previously accelerated when it passes the narrow

[3]From the Latin word *lamina*: thin sheet.

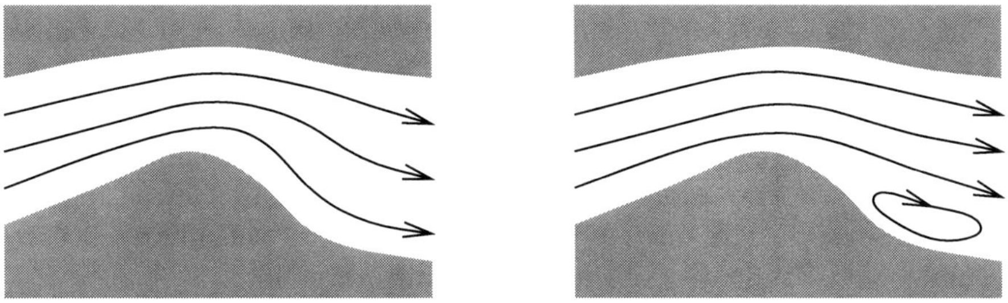

FIG. 1.3. *Vorticity generation in a river (view from above).*

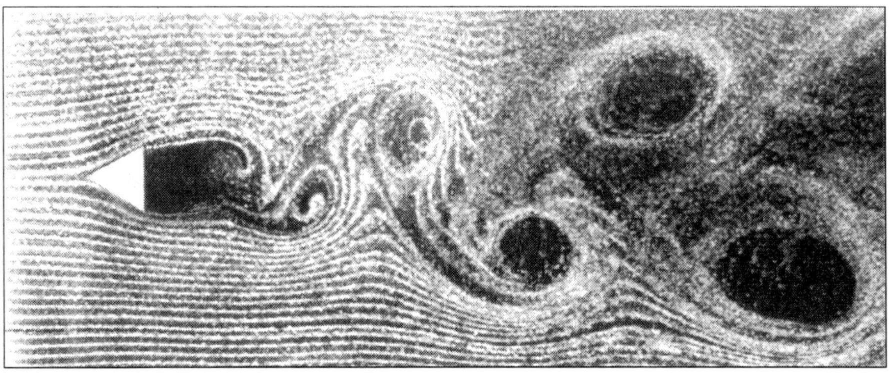

FIG. 1.4. *Kármán vortex street (from* [Nakayama, 1988]; *photograph by Y. Nakayama).*

section (Figure 1.3, left). However, if the velocity of the water is high enough for the inertial forces to dominate, then the flow separates from the bank after the narrow section, creating a vortex between the main flow and the river bank in which the flow has reversed direction (Figure 1.3, right). At such locations the fluid tends to stagnate, allowing particles carried along by the flow to be deposited. Such sediment deposition can be observed in any river.

In the flow past an obstacle, the vortices begin to separate from the obstacle once the mean flow exceeds a certain velocity, forming what is known as a *Kármán vortex street* (Figure 1.4), named after the Hungarian fluid dynamicist Theodor von Kármán. The effects of vorticity formation can be felt when a large truck passes us on the highway. When the truck is level with us, we first notice a lateral force due to the air displaced by the truck. Once the truck has passed, we notice forces pulling our car alternately to the left and to the right. These forces are the result of precisely those vortices generated by the truck.

If the velocity is yet further increased, the flow no longer remains laminar, but becomes *turbulent*. Turbulence can be observed, for instance, in the rapids of streams when the structure of the flow is no longer visible, and only foam and bubbles can be seen.[4]

Besides the distinctions between viscous and inviscid flows and laminar and

[4]The characteristic properties of flows at different velocities are described very vividly in [Trefil, 1986].

turbulent flows, one also distinguishes compressible fluids from incompressible fluids. Here, compressible means that the fluid can be compressed, i.e., that fluid of the same mass need not always occupy the same volume. The volume it occupies will depend on the pressure. Gases, in particular, are compressible at high velocities. Gases at low velocities, as well as most liquids, can be regarded as nearly incompressible. In this case, a given fluid mass will always occupy the same volume and the density of the fluid remains constant.

1.3 Numerical Flow Simulation

In this book we introduce and illustrate the typical steps of numerical simulation from modeling to visualization, employing fluid mechanics—specifically, unsteady, incompressible, laminar flow—as an example. Using the *Navier–Stokes equations*, the governing equations of fluid mechanics, one can, e.g., simulate the flow around aircraft and other vehicles in order to investigate the drag coefficient of different airfoil shapes or car body designs. In the design of buildings, the objective may be to place the chimney in such a way that emissions will not be drawn back to the house by vortices or to examine the effect a heating or air conditioning system has on the air circulation in a room. Further applications are the simulation of flood waves from a breaking dam, the simulation of liquid sloshing in a moving container, or the construction of drinking water reservoirs, so as to avoid zones of stagnant water in which the water exchange rate is no longer sufficient. But flows often play a role in more complex situations as well, e.g., weather and climate modeling and the investigation of melting and combustion processes, in which the flow equations must be augmented by equations for the variation of energy or chemical reactions among the substances involved. Another application area is concerned with the interactions between fluids and solid bodies, which can affect each other by their relative motions. An example of this is a flow deforming a membrane (e.g., in a valve) or a moving body (such as a ship) affecting the surrounding fluid. The simulation of pumps and aquifers also falls into this problem category. An overview of current simulation computations is given by, among others, [Durst, 1994] and [Krause, 1994].

The development of computational fluid dynamics (CFD) is closely linked with that of computing machinery. Although the first papers in which the numerical solution of partial differential equations was applied to problems of fluid mechanics may have appeared as early as 1933 [Thom, 1933], the emergence of numerical flow simulation as a viable alternative to physical experiments was not possible until electronic computing devices became available.

While early work in the 1940s was still much constrained by the limited capabilities of computers of that time [Crank & Nicholson, 1947], a first efficient method (ADI) for the numerical solution of elliptic and parabolic problems was introduced in the mid-1950s [Peaceman & Rachford, 1955]. But it wasn't until the mid-1960s that the potential of numerical flow simulation was properly recognized. Its capabilities were pointed out in [Harlow & Fromm, 1965], and the paper [Macagno, 1965] also attracted attention. It is widely agreed [Roache, 1976, p. 4] that the appearance of these two publications marks the beginning of numerical simulation as an independent discipline within fluid mechanics. Vigorous development of numerous methods for simulating the various flow models—for inviscid flows, boundary layer flows, compressible viscous flows, and incompressible viscous flows—subsequently occurred.

The development of numerical methods for incompressible viscous flow was strongly influenced by the *marker-and-cell* (MAC) method of Harlow et al. from the Los Alamos National Laboratory [Harlow & Welch, 1965] (the algorithms in this book are based on this method). The method consists of a simple finite difference scheme with an explicit first-order time discretization. Despite its age and simplicity, this method is surprisingly flexible and relatively efficient and, above all, easily understood. It may be applied to the computation of flows in fixed domains as well as to the simulation of free boundary value problems. These are problems in which the domain occupied by the fluid changes with time.

The years that followed brought forth many enhancements of the MAC scheme. Among these, [Nichols & Hirt, 1971] improved the free-surface boundary condition, Viecelli introduced a way to treat not only boundaries aligned with coordinate axes but also curved [Viecelli, 1969] and moving boundaries [Viecelli, 1971], and [Hirt & Cook, 1972] extended the code to the three-dimensional case. In [Daly & Pracht, 1968], multiphase flows were computed taking into account the surface tension at the interface between two different fluids. Work based on the MAC method has also continued in more recent times: Miyata and Masuko apply a modification of this method to the simulation of waves generated by ships [Miyata & Masuko, 1985] and breaking waves [Miyata, 1986], while Tome and McKee have developed a MAC variant for the simulation of the injection molding of plastics [Tome & McKee, 1994]. Moreover, techniques for treating free boundaries requiring considerably less memory have been developed in [Hirt & Nichols, 1981], [Chen et al., 1991], and [Lafaurie et al., 1994].

Numerous other methods for the computation of incompressible viscous flows were designed in later years, such as the SIMPLE-method [Patankar & Spalding, 1972], which is based on a semi-implicit time discretization, or the QUICK scheme, which uses a higher order discretization [Leonard, 1979]. Furthermore, finite volumes and finite elements on structured, block-structured, or unstructured grids were and are being applied to the discretization of the spatial derivatives in place of finite differences.

In addition to the treatment of equations in the so-called primitive variables, velocity and pressure, representation with stream function and vorticity as the unknown quantities is also used (see Section 4.3.3). The development of turbulence models of all kinds (large eddy, k-ε) has enabled the modeling and simulation of turbulent flows (see Chapter 10).

In recent years, much attention has been devoted to the development of adaptive methods in which the grid points are no longer distributed equally across the domain, the development of multigrid methods, the use of multilevel preconditioners as fast solvers for the resulting discrete systems of equations, and the parallelization of existing flow simulation codes. This is also reflected in the extensive support of research projects in the United States (Federal High Performance Computing and Communication Program (HPCC)), the European Union (Europort project), Germany (Deutsche Forschungsgesellschaft (German Research Foundation), with its program in "Flow Simulation on High Performance Computers"), and the joint French–German research program "Numerische Strömungssimulation—Simulation Numérique d'Ecoulements."

Accounts of these various methods can be found in a number of books (cf., e.g., [Roache, 1976], [Chorin & Marsden, 1979], [Peyret & Taylor, 1983], [Anderson et al., 1984], [Sod, 1985], [Fletcher, 1991]). Along with the treatment of incompressible viscous flows, these usually also discuss the numerical simulation of compressible flows, in which density becomes a new dependent variable, and inviscid flows, which are described by the Euler equations. By now, methods of all types have been implemented in a multitude of commercial and research software packages such as *PHOENICS, FLOTRAN, NSFLEX, FIDAP, FIRE, FLUENT, LiSS,* and *FASTEST*, to name only a few.

Chapter 2

The Mathematical Description of Flows

We now proceed to the treatment of *laminar* flows of *viscous*, *incompressible fluids*. In the present chapter, we introduce the appropriate mathematical model given by the *Navier–Stokes equations* along with their derivation. Moreover, we show how the physical quantities may be normalized in order to arrive at dimensionless equations. This is followed by a discussion of the *dynamic similarity* of different flows.

2.1 The Mathematical Model: The Navier–Stokes Equations

The flow of a fluid in a region $\Omega \subset \mathbb{R}^N$ $(N \in \{2, 3\})$ throughout time $t \in [0, t_{\text{end}}]$ is characterized by the following quantities:

- $\vec{u} : \Omega \times [0, t_{\text{end}}] \to \mathbb{R}^N$ velocity field,
- $p : \Omega \times [0, t_{\text{end}}] \to \mathbb{R}$ pressure,
- $\varrho : \Omega \times [0, t_{\text{end}}] \to \mathbb{R}$ density.

Incompressible flows are characterized by the property that density changes are negligible; hence we have $\varrho(\vec{x}, t) = \varrho_\infty = \text{const}$. The flow is described by a system of partial differential equations whose dimensionless form is given by[5]

$$\begin{aligned} \frac{\partial}{\partial t}\vec{u} + (\vec{u} \cdot \text{grad})\vec{u} + \text{grad}p &= \frac{1}{Re}\Delta\vec{u} + \vec{g} \quad \textit{(momentum equation)}, \\ \text{div}\vec{u} &= 0 \qquad\qquad\quad \textit{(continuity equation)}, \end{aligned} \tag{2.1}$$

[5]In two dimensions the operators div and grad are defined as follows:

$$\text{grad}u := (\partial u/\partial x, \partial u/\partial y), \qquad \text{div}\vec{u} := \partial u/\partial x + \partial v/\partial y\,.$$

For $(\vec{u} \cdot \text{grad})\vec{u}$ we obtain the componentwise expression

$$(\vec{u} \cdot \text{grad})\vec{u} = \left(\begin{pmatrix} u \\ v \end{pmatrix} \cdot \begin{pmatrix} \partial/\partial x \\ \partial/\partial y \end{pmatrix} \right) \begin{pmatrix} u \\ v \end{pmatrix} = \begin{pmatrix} u\,\partial u/\partial x + v\,\partial u/\partial y \\ u\,\partial v/\partial x + v\,\partial v/\partial y \end{pmatrix}.$$

where pressure is determined only up to an additive constant. The complete system bears the name *Navier–Stokes equations.* The quantity $Re \in \mathbb{R}$ is the dimensionless *Reynolds number*, and $\vec{g} \in \mathbb{R}^N$ denotes body forces such as gravity acting throughout the bulk of the fluid.

For simplicity, we now restrict our consideration to the two-dimensional case ($N = 2$) and rewrite equations (2.1) in component form. For $\vec{x} = (x, y)^T$, $\vec{u} = (u, v)^T$, $\vec{g} = (g_x, g_y)^T$, they then read as follows:

momentum equations:

$$\frac{\partial u}{\partial t} + \frac{\partial p}{\partial x} = \frac{1}{Re}\left(\frac{\partial^2 u}{\partial x^2} + \frac{\partial^2 u}{\partial y^2}\right) - \frac{\partial(u^2)}{\partial x} - \frac{\partial(uv)}{\partial y} + g_x\,, \tag{2.2a}$$

$$\frac{\partial v}{\partial t} + \frac{\partial p}{\partial y} = \frac{1}{Re}\left(\frac{\partial^2 v}{\partial x^2} + \frac{\partial^2 v}{\partial y^2}\right) - \frac{\partial(uv)}{\partial x} - \frac{\partial(v^2)}{\partial y} + g_y\,; \tag{2.2b}$$

continuity equation:

$$\frac{\partial u}{\partial x} + \frac{\partial v}{\partial y} = 0\,. \tag{2.2c}$$

Here, the term $(\vec{u} \cdot \text{grad})\vec{u}$ from (2.1) has been rewritten using the continuity equation.

Initially (at $t = 0$), we impose initial conditions $u = u_0(x, y)$ and $v = v_0(x, y)$ which satisfy (2.2c). In addition, conditions holding at the domain boundary are required to hold for all time, so that an *initial-boundary value problem* results.

To formulate the boundary conditions, let φ_n denote the component of velocity orthogonal to the boundary (in the exterior normal direction), φ_t the component of velocity parallel to the boundary (in the tangential direction), and $\partial\varphi_n/\partial n$ (resp., $\partial\varphi_t/\partial n$) their derivatives in the normal direction. If we restrict ourselves to domains whose boundary segments are parallel to the coordinate axes—as we will do beginning with Chapter 3—then, along the vertical boundary segments, we have

$$\varphi_n = u, \qquad \varphi_t = v, \qquad \frac{\partial \varphi_n}{\partial n} = \frac{\partial u}{\partial x}, \qquad \frac{\partial \varphi_t}{\partial n} = \frac{\partial v}{\partial x},$$

whereas, along the horizontal segments,

$$\varphi_n = v, \qquad \varphi_t = u, \qquad \frac{\partial \varphi_n}{\partial n} = \frac{\partial v}{\partial y}, \qquad \frac{\partial \varphi_t}{\partial n} = \frac{\partial u}{\partial y}.$$

When the boundary segments are not aligned with the coordinate axes, φ_n and φ_t must be calculated from u and v.

For points along the fixed boundary $\Gamma := \partial\Omega$, we shall consider the following boundary conditions.

1. *No-slip condition*: No fluid penetrates the boundary and the fluid is at rest there; i.e.,

$$\varphi_n(x, y) = 0, \qquad \varphi_t(x, y) = 0. \tag{2.3}$$

2. *Free-slip condition*: No fluid penetrates the boundary. Contrary to the no-slip condition, however, there are no frictional losses at the boundary; i.e.,

$$\varphi_n(x,y) = 0, \qquad \partial\varphi_t(x,y)/\partial n = 0. \tag{2.4}$$

   The free-slip condition is often imposed along a line or plane of symmetry in the problem, thereby reducing the size of the domain where the flow needs to be computed by a half.

3. *Inflow condition*: Both velocity components are given; i.e.,

$$\varphi_n(x,y) = \varphi_n^0, \qquad \varphi_t(x,y) = \varphi_t^0, \qquad \varphi_n^0, \varphi_t^0 \text{ given.} \tag{2.5}$$

4. *Outflow condition*: Neither velocity component changes in the direction normal to the boundary; i.e.,

$$\partial\varphi_n(x,y)/\partial n = 0, \qquad \partial\varphi_t(x,y)/\partial n = 0. \tag{2.6}$$

5. *Periodic boundary condition*: For problems which are periodic with period a in one coordinate direction (e.g., the flow over an undulating surface), one can restrict the computations to one period interval. The velocities and pressure must then coincide at the left and right boundaries; i.e.,

$$\varphi_n(0,y) = \varphi_n(a,y), \qquad \varphi_t(0,y) = \varphi_t(a,y), \qquad p(0,y) = p(a,y) \tag{2.7}$$

   (periodicity in x-direction, $x = 0$: left boundary, $x = a$: right boundary).

If the velocities, rather than their normal derivatives, are given along the entire boundary, then the boundary integral of the normal components of these given velocities must vanish;[6] i.e.,

$$\int_\Gamma \begin{pmatrix} u \\ v \end{pmatrix} \cdot \vec{n}\, ds = 0\,.$$

Examples of flows satisfying different boundary conditions are shown in Figures 2.1 and 2.2.

According to [Ladyshenskaja, 1969], the two-dimensional Navier–Stokes equations on smooth domains with smooth Dirichlet boundary conditions, i.e., with the velocities rather than their normal derivatives specified along the boundary, possess a unique solution for all time $t \geq 0$ up to an additive constant for the pressure. In three dimensions, uniqueness can be proven only for finite time intervals $[0, T]$.

In general, however, the solution or solutions cannot be calculated analytically, meaning they cannot be written down as some mathematical formula. Rather, they can most often only be approximated *numerically*.

[6] This constraint follows from the divergence theorem and the fact that the flow is divergence-free: $0 = \int_\Omega \operatorname{div}\binom{u}{v}\, dxdy = \int_\Gamma \binom{u}{v} \cdot \vec{n}\, ds$.

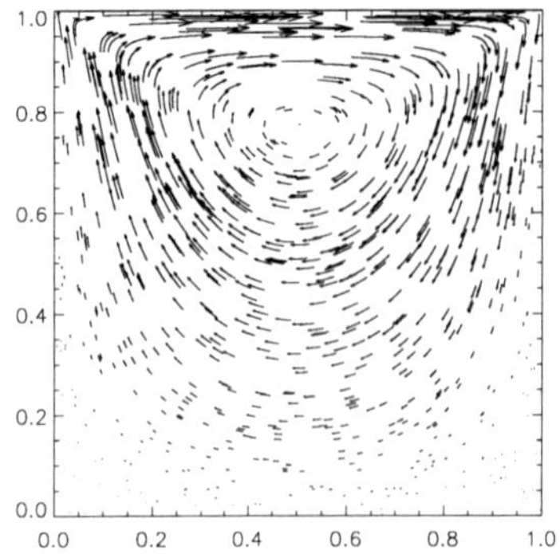

Four walls with no-slip condition.
The upper wall is moving to the right at constant speed $\varphi_t = \varphi_t^0 = \text{const.}$ along the upper boundary, in contrast with $\varphi_t = 0$ in the usual no-slip condition.

FIG. 2.1. *Driven cavity (velocity field).*

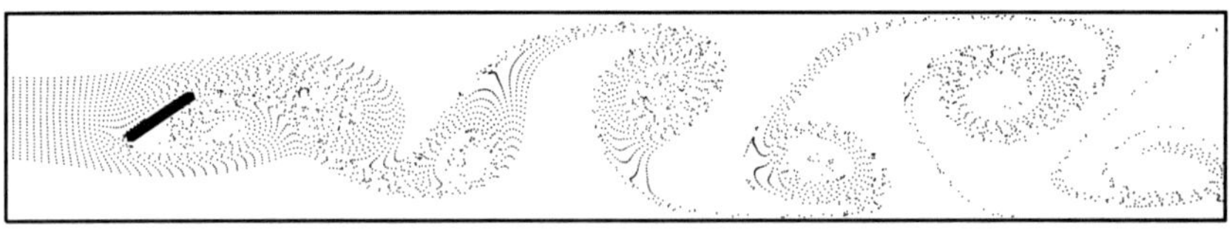

FIG. 2.2. *Flow past an inclined plate (streaklines; cf. Section 4.2)*
left: inflow, right: outflow,
top and bottom: free-slip,
no-slip on the plate.

2.2 The Derivation of the Navier–Stokes Equations

In the following, we derive the Navier–Stokes equations from the conservation laws for mass and momentum.

We consider a fluid occupying an arbitrary (open and bounded) subdomain $\Omega_0 \subset \Omega \subset \mathbb{R}^3$ of Ω at time $t = 0$. The function $\vec{\Phi} : \Omega_0 \times [0, t_{\text{end}}] \to \Omega_t \subset \Omega$ describes the motion of the particle positions $\vec{c} \in \Omega_0$ with time, so that, at time $t \geq 0$, the fluid occupies the domain $\Omega_t := \{\vec{\Phi}(\vec{c}, t) : \vec{c} \in \Omega_0\}$. Hence Ω_t is a closed system in the sense that no fluid particle flows across its boundaries (Figure 2.3).

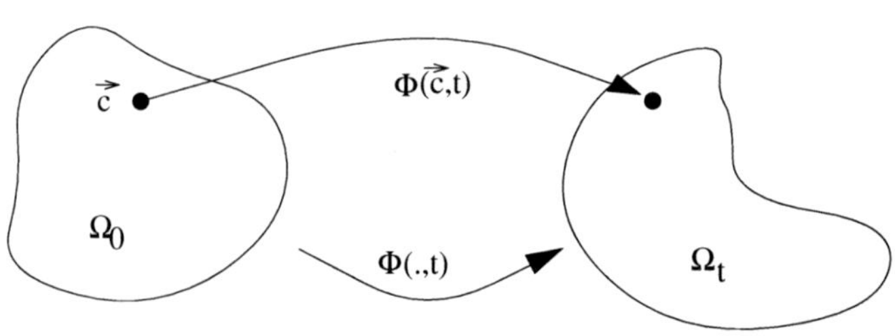

FIG. 2.3. *The closed system Ω_t.*

The path of a particle $\vec{c} \in \Omega_0$ is given by the graph of the function $t \mapsto \vec{\Phi}(\vec{c}, t)$ and the velocity of the fluid at a fixed location $\vec{x} := \vec{\Phi}(\vec{c}, t) \in \Omega_t$ by

$$\vec{u}(\vec{x}, t) = \frac{\partial}{\partial t}\vec{\Phi}(\vec{c}, t).$$

The following derivation of the governing equations relies on the *transport theorem*, which shows how the time derivative of an integral over a domain changing with time may be computed.

Transport theorem. For a differentiable scalar function $f : \Omega_t \times [0, t_{\text{end}}] \to \mathbb{R}$, $(\vec{x}, t) \mapsto f(\vec{x}, t)$, it holds that[7]

$$\frac{d}{dt}\int_{\Omega_t} f(\vec{x}, t)\, d\vec{x} = \int_{\Omega_t} \left\{ \frac{\partial}{\partial t} f + \operatorname{div}(f\vec{u}) \right\} (\vec{x}, t)\, d\vec{x}. \tag{2.8}$$

2.2.1 Conservation of Mass

The mass of a fluid occupying a domain Ω is determined by the integral over the density of the fluid. Since the same amount of fluid occupying the domain Ω_0 at time $t = 0$ later occupies the domain Ω_t at time $t > 0$, we must have for all $t \geq 0$

$$\int_{\Omega_0} \varrho(\vec{x}, 0)\, d\vec{x} = \int_{\Omega_t} \varrho(\vec{x}, t)\, d\vec{x}.$$

The derivative of mass with respect to time must therefore vanish, upon which the transport theorem (2.8) yields

$$0 = \int_{\Omega_t} \left\{ \frac{\partial}{\partial t}\varrho + \operatorname{div}(\varrho\vec{u}) \right\} (\vec{x}, t)\, d\vec{x} \qquad \forall \Omega_t,\ t \geq 0.$$

Since this is valid for arbitrary regions Ω_t (in particular, for arbitrarily small ones), this implies that the integrand itself vanishes, which yields the continuity equation for compressible fluids:

$$\frac{\partial}{\partial t}\varrho + \operatorname{div}(\varrho\vec{u}) = 0. \tag{2.9}$$

For the special case of an incompressible fluid (no density variation in time or space, i.e., $\varrho(\vec{x}, t) = \varrho_\infty = \text{const}$), this results in the continuity equation given in (2.1):

$$\operatorname{div}\vec{u} = 0. \tag{2.10}$$

[7]Since the domain Ω_t varies with time, we cannot simply differentiate under the integral sign. The derivation of this formula involves transforming the integral over Ω_t to one over Ω_0, performing the differentiation, and then transforming back from Ω_0 to Ω_t.

2.2.2 Conservation of Momentum

The momentum of a solid body is the product of its mass with its velocity. In the case of a fluid system, however, the fluid velocity may vary with position, hence the momentum of the fluid in the domain Ω_t must be expressed by the integral

$$\vec{m}(t) := \int_{\Omega_t} \varrho(\vec{x},t)\vec{u}(\vec{x},t)\,d\vec{x}.$$

According to *Newton's second law*, the time rate of change of (linear) momentum is equal to the sum of the forces acting on the fluid:

$$\frac{d}{dt}\vec{m}(t) = \sum \text{acting forces}.$$

We distinguish two types of forces:

- *body forces* (e.g., gravity, Coriolis force, magnetic force), which can be expressed as $\int_{\Omega_t} \varrho(\vec{x},t)\vec{g}(\vec{x},t)\,d\vec{x}$ with a given force density $\vec{g}$ per unit volume, and
- *surface forces* (e.g., pressure and internal friction), which may be represented as $\int_{\partial\Omega_t} \boldsymbol{\sigma}(\vec{x},t)\,\vec{n}\,ds$ in which we have introduced the *stress tensor* $\boldsymbol{\sigma}$ given by

$$\boldsymbol{\sigma} = \begin{pmatrix} \sigma_{11} & \sigma_{12} & \sigma_{13} \\ \sigma_{21} & \sigma_{22} & \sigma_{23} \\ \sigma_{31} & \sigma_{32} & \sigma_{33} \end{pmatrix}.$$

Newton's law thus reads[8]

$$\frac{d}{dt}\int_{\Omega_t} \varrho(\vec{x},t)\vec{u}(\vec{x},t)\,d\vec{x} = \int_{\Omega_t} \varrho(\vec{x},t)\,\vec{g}(\vec{x},t)\,d\vec{x} + \int_{\partial\Omega_t} \boldsymbol{\sigma}(\vec{x},t)\,\vec{n}\,ds.$$

We now apply the product rule and the transport theorem (2.8) componentwise to the term on the left and apply the divergence theorem to the first term on the right (see the footnote on page 13). We thus obtain the momentum equation[9]

$$\frac{\partial}{\partial t}(\varrho\vec{u}) + (\vec{u}\cdot\text{grad})(\varrho\vec{u}) + (\varrho\vec{u})\text{div}\vec{u} - \varrho\vec{g} - \text{div}\boldsymbol{\sigma} = 0. \tag{2.11}$$

The nature of this equation depends heavily on the model used for the stress tensor $\boldsymbol{\sigma}$. When modeling an *inviscid* fluid, internal friction is neglected and the stress tensor $\boldsymbol{\sigma}$ is determined solely by the pressure:

$$\boldsymbol{\sigma}(\vec{x},t) := -p(\vec{x},t)\mathbf{I} = -p(\vec{x},t)\begin{pmatrix} 1 & 0 & 0 \\ 0 & 1 & 0 \\ 0 & 0 & 1 \end{pmatrix}.$$

[8]Note that this is a vector equation; the integrals and time derivatives are to be understood componentwise.

[9]The divergence of a tensor is obtained by applying the divergence operator to each of its rows, resulting in a vector.

Substituting this stress tensor in the momentum equation (2.11) yields a first-order system of partial differential equations, the *Euler equations of fluid motion*:

$$\frac{\partial}{\partial t}(\varrho\vec{u}) + (\vec{u}\cdot\text{grad})(\varrho\vec{u}) + (\varrho\vec{u})\text{div}\vec{u} + \text{grad}p = \varrho\vec{g},$$

commonly used in gas dynamics (the study of compressible but inviscid flows).

The modeling of *viscous* fluids incorporates internal friction which, for *Newtonian fluids* obeying the *Stokes assumption*, results in a viscous part $\boldsymbol{\tau}$ of the stress tensor $\boldsymbol{\sigma}$:

$$\boldsymbol{\sigma} := -p\mathbf{I} + \boldsymbol{\tau} := (-p + \lambda\,\text{div}\vec{u})\mathbf{I} + 2\mu\boldsymbol{\delta}, \tag{2.12}$$

which contains the two thermodynamic material constants μ and λ as well as the *strain tensor*

$$\boldsymbol{\delta} := \frac{1}{2}\left[\left(\frac{\partial u_i}{\partial x_j} + \frac{\partial u_j}{\partial x_i}\right)\right]_{i,j=1,2,3}. \tag{2.13}$$

Substituting this stress tensor in (2.11) yields a second-order system of partial differential equations:

$$\frac{\partial}{\partial t}(\varrho\vec{u})+(\vec{u}\cdot\text{grad})(\varrho\vec{u})+(\varrho\vec{u})\text{div}\vec{u}+\text{grad}p = (\mu+\lambda)\text{grad}(\text{div}\vec{u})+\mu\Delta\vec{u}+\varrho\vec{g}\,. \tag{2.14}$$

For compressible fluids ($\varrho = \varrho(\vec{x},t)$), one requires additional information relating pressure p and density ϱ known as the *equation of state*:

$$p = r(\varrho). \tag{2.15}$$

For incompressible fluids ($\varrho(\vec{x},t) = \varrho_\infty = \text{const}$), equation (2.14) results in the momentum equation

$$\frac{\partial}{\partial t}\vec{u} + (\vec{u}\cdot\text{grad})\vec{u} + \frac{1}{\varrho_\infty}\text{grad}p = \frac{\mu}{\varrho_\infty}\Delta\vec{u} + \vec{g} \tag{2.16}$$

after using the continuity equation (2.10). Equation (2.16) contains two quantities known as the

$$\text{dynamic viscosity} \quad \mu \quad \text{and the kinematic viscosity } \nu := \frac{\mu}{\varrho_\infty}. \tag{2.17}$$

2.3 Dynamic Similarity of Flows

It is often desirable to reproduce large scale physical experiments in a scaled-down and more manageable laboratory setting. For this purpose, scale models of, e.g., aircraft wings or automobile bodies are built and the flow around these is measured in a wind tunnel. The question then arises as to what can be inferred about the full size aircraft wing or automobile body from observations collected using the models (see Figure 2.4).

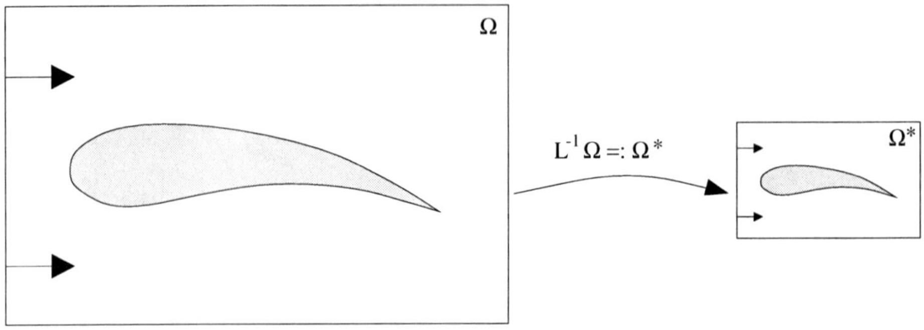

FIG. 2.4. *Reduction of the large model by the factor* $L^{-1} < 1$: *the geometries* Ω *and* $\Omega^*(= L^{-1} \cdot \Omega)$ *are* similar.

Information on the flow is contained in the parameters which characterize it, such as the dynamic viscosity μ and characteristic values for the length L, the velocity u_∞, and the density ϱ_∞. If these parameters are combined in a suitable way to yield dimensionless quantities, then these enable one to make the desired statements relating the flows on the large and small scales. We thus form dimensionless quantities from their dimensional counterparts in the following way:

$$\text{dimensionless quantity} = \frac{\text{dimensional quantity}}{\text{reference quantity with the same physical unit}}.$$

The reference quantities used should possess certain properties; specifically, they should be

- constant for the problem,
- known in advance, and
- characteristic for the problem.

We turn our attention to the Navier–Stokes equations for incompressible flows (2.16) and introduce the dimensionless variables

$$\vec{x}^* := \frac{\vec{x}}{L}, \quad t^* := \frac{u_\infty t}{L}, \quad \vec{u}^* := \frac{\vec{u}}{u_\infty}, \quad p^* := \frac{p - p_\infty}{\varrho_\infty u_\infty^2} \tag{2.18}$$

with given scalar constants L, u_∞, p_∞, ϱ_∞.[10]

Recasting equation (2.16) in the variables (2.18) leads to the equation

$$\frac{\partial}{\partial t^*}\vec{u}^* + (\vec{u}^* \cdot \text{grad}^*)\vec{u}^* + \text{grad}^* p^* = \frac{\mu}{\varrho_\infty u_\infty L}\Delta^* \vec{u}^* + \frac{L}{u_\infty^2}\vec{g}, \tag{2.19}$$

in which the operators grad^* and Δ^* are those with respect to the variable $\vec{x}^*$ rather than $\vec{x}$. Since this equation depends only on the parameter groupings on

[10] In a wind tunnel, these could be the length L of the obstacle in the flow and u_∞, p_∞, and ϱ_∞ the upstream values of velocity, pressure, and density.

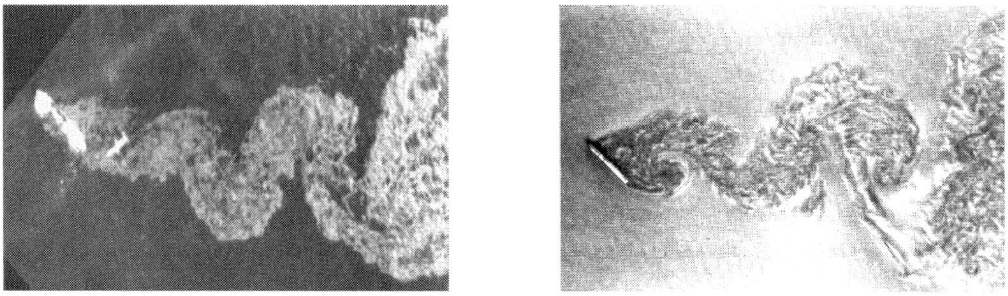

FIG. 2.5. *Aerial photo of tanker accident (left) and laboratory experiment in glycerin (right) (from* [Cantwell, 1981]*).*

the right-hand side, two flows will behave identically whenever their parameters are such that both parameter groupings coincide. This brings us to the following conclusion:

Flows in similar geometries $\Omega^* = L^{-1} \cdot \Omega$, $L > 0$, are *dynamically similar*, if the parameters μ, u_∞, ϱ_∞, L, $\vec{g}$ of each flow are such that the dimensionless quantities[11]

$$Re := \frac{\varrho_\infty u_\infty L}{\mu} \quad \text{(Reynolds number)} \qquad \text{and} \qquad Fr := \frac{u_\infty}{\sqrt{L\,\|\vec{g}\|}} \quad \text{(Froude number)} \tag{2.20}$$

of the flows coincide.

By introducing the dimensionless body force

$$\vec{g}^* := \frac{L}{u_\infty^2}\vec{g} \left(= \frac{1}{Fr^2}\frac{\vec{g}}{\|\vec{g}\|} \right),$$

we obtain the momentum equation in (2.1) from (2.19).

The dimensionless Reynolds number Re and Froude number Fr both describe properties of the flow. The Reynolds number represents the relative magnitude of inertial and viscous forces: for $Re \approx 0$ the inertial forces are negligible against the viscous forces (highly viscous fluid), whereas for Re very large (up to 10^{10} or larger) the viscous forces can be neglected (e.g., for air). The Froude number measures the ratio of inertial forces to gravitational forces.

Figure 2.5 shows an aerial photo of an oil tanker accident on the open sea next to a small scale laboratory experiment performed in glycerin to reproduce this flow. One clearly recognizes the similarity of the two flows. An example for the conversion of dimensional quantities to nondimensional ones is given in Section 9.7 on page 141, where temperature is also included in the model.

[11] $\|\vec{g}\|$ denotes the (euclidean) length of the vector $\vec{g}$: $\|\vec{g}\| := \sqrt{g_x^2 + g_y^2}$ in two dimensions and $\sqrt{g_x^2 + g_y^2 + g_z^2}$ in three dimensions, respectively.

In practice, however, successfully exploiting the relationship among dynamically similar flows often requires much effort. If an airplane model reduced by a factor of 30 is studied in a wind tunnel, the resulting flow will have a Reynolds number 30 times smaller than for the real airplane. Hence, for engineers to be able to draw conclusions on the flight behavior of the full size airplane, conditions in the wind tunnel must be adjusted accordingly. For conventional wind tunnels using air as their *wind tunnel gas*, this is usually attempted by increasing the air pressure in the experimental chamber. This is generally not sufficient, and hence many wind tunnel tests are performed at Reynolds numbers only one-fifth of the desired magnitude.

In a recent European project, the European Transsonic Wind Tunnel (ETW), a new type of wind tunnel was constructed in Cologne-Porz, Germany. Besides increasing the pressure, it uses two additional techniques in order to attain the correct Reynolds number for wind tunnel experiments: nitrogen is used instead of air as the wind tunnel gas, which in addition is cooled down to -183° Celsius.[12] This technical effort may have its price (660 million DM, or 390 million dollars, for the construction of the facility and power consumption of 50 megawatts, which corresponds to that of a small town), but it now permits wind tunnel testing under absolutely realistic conditions.

[12]Using air at such low temperatures is out of the question, as the oxygen it contains would liquefy.

Chapter 3

The Numerical Treatment of the Navier–Stokes Equations

In this chapter we describe the numerical solution of the unsteady incompressible Navier–Stokes equations. We introduce the method of *finite differences* for the discretization of simple differential equations and apply it to the continuous Navier–Stokes equations, resulting in a finite-dimensional (discrete) problem. For the discretized Navier–Stokes equations, we give detailed instructions on the implementation of a computer program for flow calculations on rectangular domains. We conclude by introducing a method for extending the rectangular computational domain to more general two-dimensional domains.

3.1 The Discretization

In numerical analysis, the term *discretization* refers to passing from a continuous problem to one considered at only a finite number of points. In particular, discretization is used in the numerical solution of differential equations by reducing the differential equation to a system of algebraic equations. These determine the values of the solution at only a finite number of points of the domain on which the solution is sought. In what follows, we will use the *finite difference method.*[13]

3.1.1 Simple Discretization Formulas

Discretization in One Dimension

We begin with the one-dimensional case. The interval $\Omega := [0, a] \subset \mathbb{R}$, on which a differential equation is to be solved, is subdivided into $i_{\max}$ subintervals of equal size, $\delta x := a/i_{\max}$. We thus obtain a grid consisting of the points $x_i := i\,\delta x$, $i = 0, \ldots, i_{\max}$, at the subinterval boundaries (see Figure 3.1).

[13]There are many other discretization methods, such as the finite element method (cf. [Ciarlet, 1978], [Strang & Fix, 1973], [Brenner & Scott, 1994]), the finite volume method, also known as the box method (cf. [Patankar, 1980]), and the class of spectral methods (cf. [Canuto et al., 1988]).

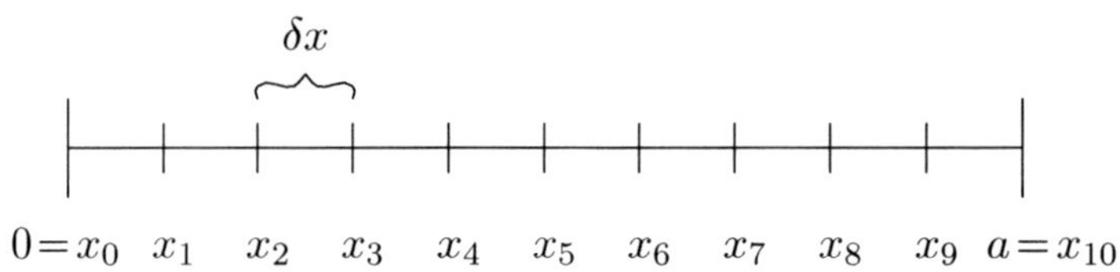

FIG. 3.1. *One-dimensional equispaced grid with* $i_{\max} = 10$.

The differential equation is now considered only at these grid points. Recalling the definition of the derivative

$$\frac{du}{dx} := \lim_{\delta x \to 0} \frac{u(x+\delta x) - u(x)}{\delta x} \tag{3.1}$$

of a differentiable function $u : \mathbb{R} \to \mathbb{R}$, we now approximate the continuous differential operator du/dx at a grid point x_i by the discrete difference operator

$$\left[\frac{du}{dx}\right]_i^r := \frac{u(x_{i+1}) - u(x_i)}{\delta x} \tag{3.2}$$

by not passing to the limit. Here $x_{i+1} = x_i + \delta x$ is the right neighboring grid point of x_i. Besides this so-called *forward difference*, one can also form the *backward difference*

$$\left[\frac{du}{dx}\right]_i^l := \frac{u(x_i) - u(x_{i-1})}{\delta x} \tag{3.3}$$

and the *central difference*

$$\left[\frac{du}{dx}\right]_i^c := \frac{u(x_{i+1}) - u(x_{i-1})}{2\,\delta x}. \tag{3.4}$$

It is intuitively clear that refining the grid, i.e., reducing the stepsize δx, leads to a better approximation of the differential quotient. In fact, the discretization error of the forward and backward differences is of the order $O(\delta x)$, which means that by halving the stepsize one may expect the error to decrease by one-half as well. However, if one instead uses the central difference for the approximation, the error is of the order $O(\delta x)^2$, so that halving the stepsize roughly reduces the error by a factor of four. Note that, for these statements to hold, the function u must be sufficiently smooth.

To approximate the second derivative d^2u/dx^2 of a function u at a grid point x_i, we form the central differences of the first derivatives at the points $x_{i+1/2} := x_i + \delta x/2$ and $x_{i-1/2} := x_i - \delta x/2$ using half the stepsize. These first derivatives are in turn approximated using central differences with half the stepsize. Thus we approximate the derivative of the derivative of u as follows:

$$\left[\frac{d^2u}{dx^2}\right]_i := \frac{1}{\delta x}\left(\left[\frac{du}{dx}\right]_{i+\frac{1}{2}}^c - \left[\frac{du}{dx}\right]_{i-\frac{1}{2}}^c\right) = \frac{u(x_{i+1}) - 2\,u(x_i) + u(x_{i-1})}{\delta x^2}. \tag{3.5}$$

Given an ordinary differential equation of second order,[14]

$$-\frac{d^2u}{dx^2} + k\,\frac{du}{dx} = f \quad \text{in } \Omega, \tag{3.6}$$

with so-called Dirichlet boundary conditions

$$u(0) =: u_0, \quad u(a) =: u_{i_{\max}},$$

we can replace the derivatives with the appropriate difference quotients from (3.4) and (3.5) at each grid point x_i, $i = 1, \ldots, i_{\max} - 1$, to obtain a system of $i_{\max} - 1$ equations

$$\frac{1}{\delta x^2}\left(-u_{i+1} + 2\,u_i - u_{i-1}\right) + \frac{k}{2\,\delta x}\left(u_{i+1} - u_{i-1}\right) = f(x_i)\,, \quad i = 1, \ldots, i_{\max} - 1, \tag{3.7}$$

with unknowns u_i approximating the function values of u at the interior grid points x_i. In matrix-vector notation, this reads

$$A\,u = f \tag{3.8}$$

with

$$A = \begin{pmatrix} a_{1,1} & \cdots & a_{1,i_{\max}-1} \\ \vdots & & \vdots \\ a_{i_{\max}-1,1} & \cdots & a_{i_{\max}-1,i_{\max}-1} \end{pmatrix}, \quad u = \begin{pmatrix} u_1 \\ \vdots \\ u_{i_{\max}-1} \end{pmatrix}, \quad f = \begin{pmatrix} f(x_1) \\ \vdots \\ f(x_{i_{\max}-1}) \end{pmatrix},$$

where the entries of the matrix A are defined by

$$a_{i,i} := \frac{2}{\delta x^2}, \; a_{i,i-1} := -\frac{k}{2\delta x} - \frac{1}{\delta x^2}, \; a_{i,i+1} := +\frac{k}{2\delta x} - \frac{1}{\delta x^2}, \; a_{i,j} := 0 \text{ for } |i-j| > 1.$$

When convection-diffusion problems with dominating convection term (k large) are discretized using central differences, stability problems occur when the grid spacing δx is chosen too coarse. This results in unphysical oscillations in the numerical solution (see Figure 3.2). The reason for this lies in the fact that, for δx too large, certain properties of the continuous equation are no longer correctly captured by the discrete equation. The eigenvalues of the matrix A, for instance, should all have positive real parts, but this is violated as soon as

$$\delta x > \frac{2}{|k|}.$$

Maintaining stability for strongly convective problems would thus restrict the grid spacing to be very small, leading to excessively large systems of equations, particularly in two and three dimensions.

[14] Equations of this form are called *convection-diffusion equations*. They can be used to describe, e.g., the temperature distribution in a moving fluid (cf. also Chapter 9). By molecular diffusion, heat spreads out uniformly in all directions; this process is described by the term d^2u/dx^2. The second term, $k\,du/dx$, on the other hand, describes the transport of heat with the flow (of velocity k), a process known as *convection*.

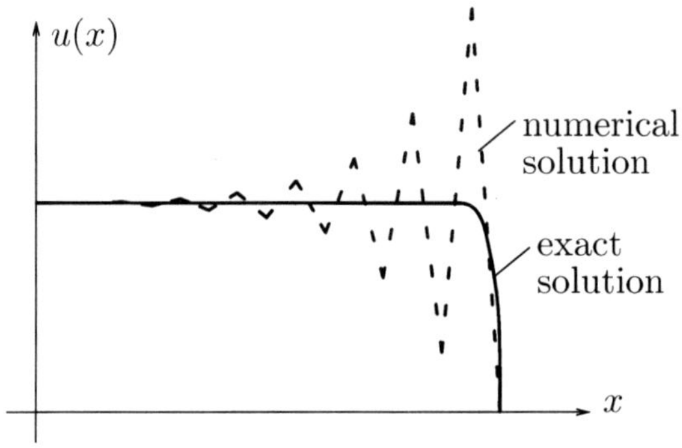

FIG. 3.2. *Oscillations in convection-diffusion problems when central differences are used.*

If we replace the central difference approximating du/dx by one-sided differences, i.e., a backward difference for positive k and a forward difference for negative k, viz.,

$$\left[\frac{du}{dx}\right]_i^{\mathrm{up}} := \frac{(1+\epsilon)(u_i - u_{i-1}) + (1-\epsilon)(u_{i+1} - u_i)}{2\,\delta x} \quad \text{with} \quad \epsilon := \mathrm{sign}(k), \qquad (3.9)$$

then all eigenvalues of the system matrix A have positive real part for all choices of δx. The price to pay for this stable discretization known as *upwind differencing* or simply *upwinding* is a drop in the order of approximation from $O(\delta x^2)$ to $O(\delta x)$ (cf. [Hackbusch, 1992, Chapter 10.2.2]).

A possible compromise lies in using a weighted average of both discretizations

$$\gamma \cdot \text{upwind difference} + (1-\gamma) \cdot \text{central difference}. \qquad (3.10)$$

The real parameter γ must be chosen from the interval $[0, 1]$ and must increase with the convection parameter k.

Another possibility for avoiding stability problems is discretizing with the *donor-cell scheme* (cf. [Gentry et al., 1966]), although it also suffers from a lower order of approximation. It is applied mainly to the discretization of convection terms of the form $d(ku)/dx$. Let u be given at the grid points, but assume that k is given at the interval midpoints (see Figure 3.3).

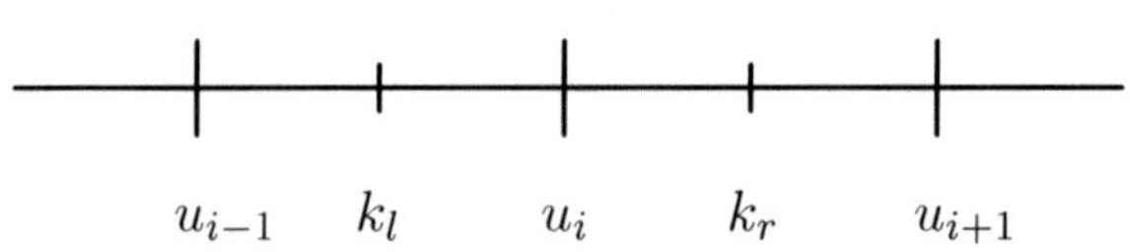

FIG. 3.3. *Discretization using the donor-cell scheme.*

The donor-cell discretization of the convection term at the grid point x_i is then given by

$$\left[\frac{d(ku)}{dx}\right]_i^{\mathrm{dc}} := \frac{k_r u_r - k_l u_l}{\delta x}, \qquad (3.11)$$

where u_r and u_l are chosen depending on the sign of k_r and k_l:

$$u_r := \begin{cases} u_i, & k_r > 0, \\ u_{i+1}, & k_r < 0, \end{cases} \qquad u_l := \begin{cases} u_{i-1}, & k_l > 0, \\ u_i, & k_l < 0. \end{cases}$$

To avoid these case distinctions, (3.11) can also be written as

$$\begin{aligned} \left[\frac{d(ku)}{dx}\right]_i^{\mathrm{dc}} &= \frac{1}{2\delta x} && ((k_r - |k_r|)\, u_{i+1} \\ & && + (k_r + |k_r| - k_l + |k_l|)\, u_i + (-k_l - |k_l|)\, u_{i-1}) \\ &= \frac{1}{2\delta x} && (k_r(u_i + u_{i+1}) - k_l(u_{i-1} + u_i) \\ & && + |k_r|(u_i - u_{i+1}) - |k_l|(u_{i-1} - u_i)). \end{aligned} \tag{3.12}$$

Thus, if the sign of k denotes the flow direction, this scheme always selects the value which lies in the upstream direction. An averaging of central and donor-cell differences similar to (3.10) is another option.[15]

Discretization in Two Dimensions

In two dimensions, we initially restrict ourselves to a rectangular region

$$\Omega := [0, a] \times [0, b] \subset \mathbb{R}^2$$

on which we introduce a grid. This grid is divided into $i_{\max}$ cells of equal size in the x-direction and $j_{\max}$ cells in the y-direction, resulting in grid lines spaced at a distance

$$\delta x := \frac{a}{i_{\max}} \quad \text{and} \quad \delta y := \frac{b}{j_{\max}}$$

apart.[16] The problem to be solved is now only considered at the intersection points of the grid lines $x_{i,j} := (i\,\delta x, j\,dy)$, $i = 0, \ldots, i_{\max}$, $j = 0, \ldots, j_{\max}$. Using (3.5), the Laplace operator

$$\Delta u := \frac{\partial^2 u}{\partial x^2} + \frac{\partial^2 u}{\partial y^2}$$

is discretized at the grid point $x_{i,j}$ as follows:

---

[15] There are numerous other methods for treating convection-dominated problems such as the finite difference technique described in [Il'in, 1969] or the streamline-diffusion [Johnson, 1987], streamline-upwind-Petrov–Galerkin, and Galerkin-least-squares [Hughes et al., 1986] methods for finite elements.

[16] To better approximate the solution of the continuous problem it is often advisable to locally reduce the grid line spacing in those subregions where the solution changes rapidly. When the solution behavior is roughly known in advance (boundary layers), such a refinement may be performed at the outset of the discretization procedure. Yet there are also other methods which employ error estimators to adaptively refine the grid during the solution process [Babuška et al., 1986], [Bank, 1994], [Johnson, 1987], [McCormick, 1989].

$$[\Delta u]_{i,j} := \frac{u(x_{i+1,j}) - 2\,u(x_{i,j}) + u(x_{i-1,j})}{\delta x^2} + \frac{u(x_{i,j+1}) - 2\,u(x_{i,j}) + u(x_{i,j-1})}{\delta y^2}. \tag{3.13}$$

The discretization of Poisson's equation with Dirichlet boundary conditions

$$\Delta u = f \quad \text{in} \quad \Omega, \qquad u = g \quad \text{on} \quad \partial\Omega \tag{3.14}$$

then leads to the linear system of equations

$$\frac{1}{\delta x^2}\left(u_{i+1,j} - 2\,u_{i,j} + u_{i-1,j}\right) + \frac{1}{\delta y^2}\left(u_{i,j+1} - 2\,u_{i,j} + u_{i,j-1}\right) = f(x_{i,j}), \tag{3.15}$$

$$i = 1, \ldots, i_{\max} - 1, \quad j = 1, \ldots, j_{\max} - 1,$$

of dimension $(i_{\max} - 1)(j_{\max} - 1)$ with unknowns $u_{i,j}$ approximating the solution at the grid points $x_{i,j}$. To satisfy the boundary condition we set $u_{i,j} = g(x_{i,j})$ for $i \in \{0, i_{\max}\}$ or $j \in \{0, j_{\max}\}$.

If the differential equation also contains derivatives of first order, then these can lead to stability problems like those we encountered in the one-dimensional case. These can again be circumvented by using upwind or donor cell discretizations.

3.1.2 Discretization of the Navier–Stokes Equations

Treatment of the Spatial Derivatives

When solving the Navier-Stokes equations, the region Ω is often discretized using a *staggered grid*, in which the different unknown variables are not located at the same grid points. In the grid we shall use, the pressure p is located in the cell centers, the horizontal velocity u in the midpoints of the vertical cell edges, and the vertical velocity v in the midpoints of the horizontal cell edges.[17] Cell (i, j) occupies the spatial region $[(i-1)\,\delta x, i\,\delta x] \times [(j-1)\,\delta y, j\,\delta y]$, and the corresponding index (i, j) is assigned to the pressure at the cell center as well as to the u-velocity at the right edge and the v-velocity at the upper edge of this cell. Hence the pressure value $p_{i,j}$ is located at the coordinates $((i - 0.5)\,\delta x, (j - 0.5)\,\delta y)$, the horizontal velocity value $u_{i,j}$ at the coordinates $(i\,\delta x, (j - 0.5)\,\delta y)$, and the vertical velocity value $v_{i,j}$ at the coordinates $((i - 0.5)\,\delta x, j\,\delta y)$ (see Figure 3.4).

As a result, the discrete values of u, v, and p are actually located on three separate grids not shown here, each shifted by half a grid spacing to the bottom, to the left, and to the lower left, respectively.

Consequently, not all extremal grid points come to lie on the domain boundary. The vertical boundaries, for instance, carry no v-values, just as the horizontal boundaries carry no u-values. For this reason, an extra boundary strip of grid cells is introduced (see Figure 3.5), so that the boundary conditions may be applied by averaging the nearest grid points on either side (see the subsection "Boundary Values for the Discrete Equations" on page 30).

[17]The idea for this staggered arrangement of the variables comes from the finite volume method, in which the continuity equation is discretized in each volume cell by considering the mass flux across the cell edges determined by the velocities on these edges.

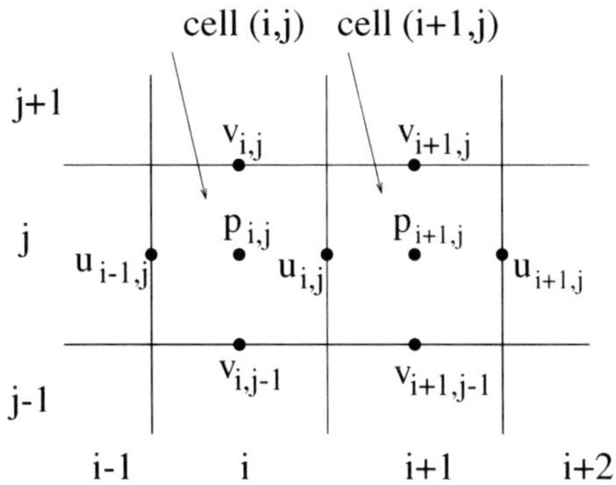

FIG. 3.4. *Staggered grid.*

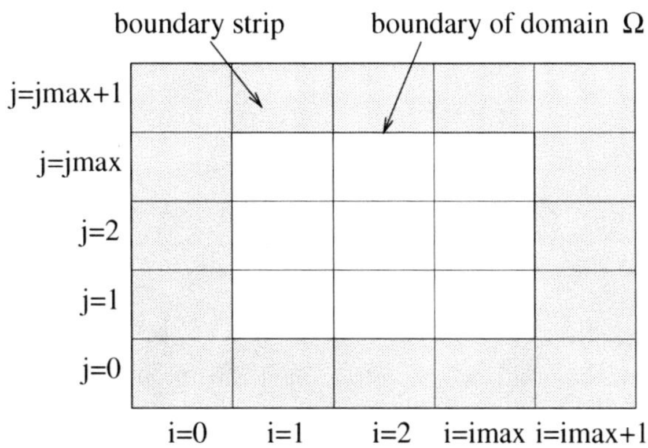

FIG. 3.5. *Domain with boundary cells.*

This staggered arrangement of the unknowns prevents possible pressure oscillations which could occur had we evaluated all three unknown functions u, v, and p at the same grid points. Consider briefly the problem with zero Dirichlet conditions imposed for both u and v along the entire boundary and $g_x = g_y = 0$. The solution of the continuous problem in this case is

$$u = 0, \qquad v = 0, \qquad p = \text{const.}$$

Were we to discretize the pressure derivative term in the momentum equations using central differences on a nonstaggered grid, then

$$u_{i,j} = 0, \qquad v_{i,j} = 0, \qquad p_{i,j} = P_1 \text{ for } i+j \text{ even}, \qquad p_{i,j} = P_2 \text{ for } i+j \text{ odd}$$

would solve the discrete problem for arbitrary values of P_1 and P_2.[18]

One alternative to a staggered grid is the use of so-called *colocated grids* (cf. [Perić et al., 1988]), for which u, v, and p are all evaluated at cell centers and a finite volume-based discretization employs special interpolation schemes to determine the flux across the cell edges.

[18]This phenomenon also occurs in finite element discretizations whenever the Babuška–Brezzi condition [Brezzi, 1974], which insures uniqueness and stability of the solution of saddlepoint problems, is violated.

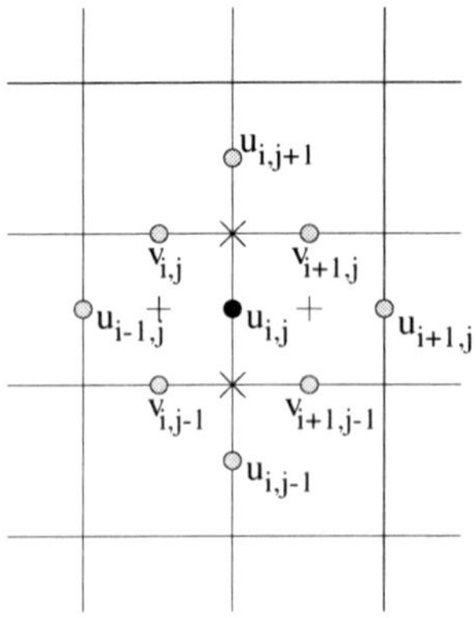

FIG. 3.6. *Values required for the discretization of the u-momentum equation.*

We return to the discretization on the staggered grid. The continuity equation (2.2c) is discretized at the center of each cell (i,j), $i = 1, \ldots, i_{\max}$, $j = 1 \ldots, j_{\max}$, by replacing the spatial derivatives $\partial u/\partial x$ and $\partial v/\partial y$ by centered differences using half the mesh width:

$$\left[\frac{\partial u}{\partial x}\right]_{i,j} := \frac{u_{i,j} - u_{i-1,j}}{\delta x}, \qquad \left[\frac{\partial v}{\partial y}\right]_{i,j} := \frac{v_{i,j} - v_{i,j-1}}{\delta y}. \tag{3.16}$$

The momentum equation (2.2a) for u, on the other hand, is discretized at the midpoints of the vertical cell edges, and the momentum equation (2.2b) for v at the midpoints of the horizontal cell edges.

The second derivatives $\partial^2 u/\partial x^2$, $\partial^2 u/\partial y^2$, $\partial^2 v/\partial x^2$, and $\partial^2 v/\partial y^2$ forming the so-called diffusive terms, in turn, can be replaced by their discrete counterparts according to (3.5) while the spatial derivatives of pressure are again treated using central differences with half the mesh width.

The discretization of the convective terms $\partial(u^2)/\partial x$, $\partial(uv)/\partial y$, $\partial(uv)/\partial x$, and $\partial(v^2)/\partial y$, however, poses some difficulties. For example, to discretize $\partial(uv)/\partial y$ at the midpoint of the right edge of cell (i,j) (black dot in Figure 3.6), we need suitable values of the product uv lying in the two vertical directions. One solution that suggests itself here is to use averages of u and v taken at the locations marked with an $\times$ in Figure 3.6, which gives us the discrete term

$$\begin{aligned}\left[\frac{\partial(uv)}{\partial y}\right]_{i,j} := \frac{1}{\delta y}\Bigg(&\frac{(v_{i,j} + v_{i+1,j})}{2}\frac{(u_{i,j} + u_{i,j+1})}{2} \\ &- \frac{(v_{i,j-1} + v_{i+1,j-1})}{2}\frac{(u_{i,j-1} + u_{i,j})}{2}\Bigg).\end{aligned} \tag{3.17}$$

Similarly, to discretize $\partial(u^2)/\partial x$, we use a central difference with half the mesh width of values averaged at the points marked with a $+$ in Figure 3.6, rather than employing a central difference between $u_{i+1,j}$ and $u_{i-1,j}$:

$$\left[\frac{\partial(u^2)}{\partial x}\right]_{i,j} := \frac{1}{\delta x}\left(\left(\frac{u_{i,j} + u_{i+1,j}}{2}\right)^2 - \left(\frac{u_{i-1,j} + u_{i,j}}{2}\right)^2\right). \tag{3.18}$$

Because the convective terms in the momentum equations become dominant at high Reynolds numbers or high velocities,[19] it is necessary to use a mixture of the central differences described above and the donor-cell discretization. Following (3.12), we set

$$k_r := \frac{u_{i,j} + u_{i+1,j}}{2}, \qquad k_l := \frac{u_{i-1,j} + u_{i,j}}{2}$$

in the donor-cell term to discretize $\partial(u^2)/\partial x$ as well as

$$k_r := \frac{v_{i,j} + v_{i+1,j}}{2}, \qquad k_l := \frac{v_{i,j-1} + v_{i+1,j-1}}{2}$$

for the discretization of $\partial(uv)/\partial y$. The remaining two terms $\partial(uv)/\partial x$ and $\partial(v^2)/\partial y$ of this type are treated analogously. In sum, we obtain the following discrete expressions (cf. [Hirt et al., 1975]).

In equation (2.2a) for u at the midpoint of the *right edge* of cell (i, j), $i = 1, \ldots, i_{\max} - 1$, $j = 1 \ldots, j_{\max}$, we set

$$\begin{aligned}
\left[\frac{\partial(u^2)}{\partial x}\right]_{i,j} &:= \frac{1}{\delta x}\left(\left(\frac{u_{i,j} + u_{i+1,j}}{2}\right)^2 - \left(\frac{u_{i-1,j} + u_{i,j}}{2}\right)^2\right) \\
&+ \gamma\frac{1}{\delta x}\left(\frac{|u_{i,j} + u_{i+1,j}|}{2}\frac{(u_{i,j} - u_{i+1,j})}{2} - \frac{|u_{i-1,j} + u_{i,j}|}{2}\frac{(u_{i-1,j} - u_{i,j})}{2}\right), \\
\left[\frac{\partial(uv)}{\partial y}\right]_{i,j} &:= \frac{1}{\delta y}\left(\frac{(v_{i,j} + v_{i+1,j})}{2}\frac{(u_{i,j} + u_{i,j+1})}{2} - \frac{(v_{i,j-1} + v_{i+1,j-1})}{2}\frac{(u_{i,j-1} + u_{i,j})}{2}\right) \\
&+ \gamma\frac{1}{\delta y}\left(\frac{|v_{i,j} + v_{i+1,j}|}{2}\frac{(u_{i,j} - u_{i,j+1})}{2} - \frac{|v_{i,j-1} + v_{i+1,j-1}|}{2}\frac{(u_{i,j-1} - u_{i,j})}{2}\right), \\
\left[\frac{\partial^2 u}{\partial x^2}\right]_{i,j} &:= \frac{u_{i+1,j} - 2u_{i,j} + u_{i-1,j}}{(\delta x)^2}, \\
\left[\frac{\partial^2 u}{\partial y^2}\right]_{i,j} &:= \frac{u_{i,j+1} - 2u_{i,j} + u_{i,j-1}}{(\delta y)^2}, \qquad \left[\frac{\partial p}{\partial x}\right]_{i,j} := \frac{p_{i+1,j} - p_{i,j}}{\delta x},
\end{aligned} \tag{3.19a}$$

and in equation (2.2b) for v at the midpoint of the *upper edge* of cell (i, j), $i = 1, \ldots, i_{\max}$, $j = 1 \ldots, j_{\max} - 1$, we set

$$\begin{aligned}
\left[\frac{\partial(uv)}{\partial x}\right]_{i,j} &:= \frac{1}{\delta x}\left(\frac{(u_{i,j} + u_{i,j+1})}{2}\frac{(v_{i,j} + v_{i+1,j})}{2} - \frac{(u_{i-1,j} + u_{i-1,j+1})}{2}\frac{(v_{i-1,j} + v_{i,j})}{2}\right) \\
&+ \gamma\frac{1}{\delta x}\left(\frac{|u_{i,j} + u_{i,j+1}|}{2}\frac{(v_{i,j} - v_{i+1,j})}{2} - \frac{|u_{i-1,j} + u_{i-1,j+1}|}{2}\frac{(v_{i-1,j} - v_{i,j})}{2}\right), \\
\left[\frac{\partial(v^2)}{\partial y}\right]_{i,j} &:= \frac{1}{\delta y}\left(\left(\frac{v_{i,j} + v_{i,j+1}}{2}\right)^2 - \left(\frac{v_{i,j-1} + v_{i,j}}{2}\right)^2\right) \\
&+ \gamma\frac{1}{\delta y}\left(\frac{|v_{i,j} + v_{i,j+1}|}{2}\frac{(v_{i,j} - v_{i,j+1})}{2} - \frac{|v_{i,j-1} + v_{i,j}|}{2}\frac{(v_{i,j-1} - v_{i,j})}{2}\right),
\end{aligned}$$

[19]Due to the identity $\partial(u^2)/\partial x = 2u\,\partial u/\partial x$, u enters as a factor in front of the first derivative $\partial u/\partial x$ in the convective terms.

$$\left[\frac{\partial^2 v}{\partial x^2}\right]_{i,j} := \frac{v_{i+1,j} - 2v_{i,j} + v_{i-1,j}}{(\delta x)^2},$$

$$\left[\frac{\partial^2 v}{\partial y^2}\right]_{i,j} := \frac{v_{i,j+1} - 2v_{i,j} + v_{i,j-1}}{(\delta y)^2}, \qquad \left[\frac{\partial p}{\partial y}\right]_{i,j} := \frac{p_{i,j+1} - p_{i,j}}{\delta y}. \tag{3.19c}$$

The parameter γ in the above formulas lies between 0 and 1. For $\gamma = 0$ we recover the central difference discretization, and for $\gamma = 1$, a pure donor-cell scheme results. According to [Hirt et al., 1975], γ should be chosen such that

$$\gamma \geq \max_{i,j}\left(\left|\frac{u_{i,j}\delta t}{\delta x}\right|, \left|\frac{v_{i,j}\delta t}{\delta y}\right|\right) \tag{3.20}$$

is satisfied.

Boundary Values for the Discrete Equations

The discretization (3.19a) of the momentum equation (2.2a) for u involves u-values on the boundary for $i \in \{1, i_{\max} - 1\}$. Moreover, for $j \in \{1, j_{\max}\}$, v-values lying on the boundary are required as well as additional u-values lying outside the domain Ω. Similarly, boundary values of v are required in the discretization (3.19c) of the momentum equation (2.2b) for v. In total, we require the values

$$\begin{aligned} &u_{0,j}, \quad u_{i_{\max},j}, \quad j = 1, \dots, j_{\max}, \\ &v_{i,0}, \quad v_{i,j_{\max}}, \quad i = 1, \dots, i_{\max}, \end{aligned}$$

on the boundary as well as the values

$$\begin{aligned} &u_{i,0}, \quad u_{i,j_{\max}+1}, \quad i = 1, \dots, i_{\max}, \\ &v_{0,j}, \quad v_{i_{\max}+1,j}, \quad j = 1, \dots, j_{\max}, \end{aligned}$$

outside the domain Ω. These velocity values are obtained from a discretization of the boundary conditions of the continuous problem.

1. *No-slip condition:* The continuous velocities should vanish at the boundary to satisfy the no-slip condition.[20] For the values lying directly on the boundary we thus set

$$\begin{aligned} &u_{0,j} = 0, \quad u_{i_{\max},j} = 0, \quad j = 1, \dots, j_{\max}, \\ &v_{i,0} = 0, \quad v_{i,j_{\max}} = 0, \quad i = 1, \dots, i_{\max}. \end{aligned} \tag{3.21}$$

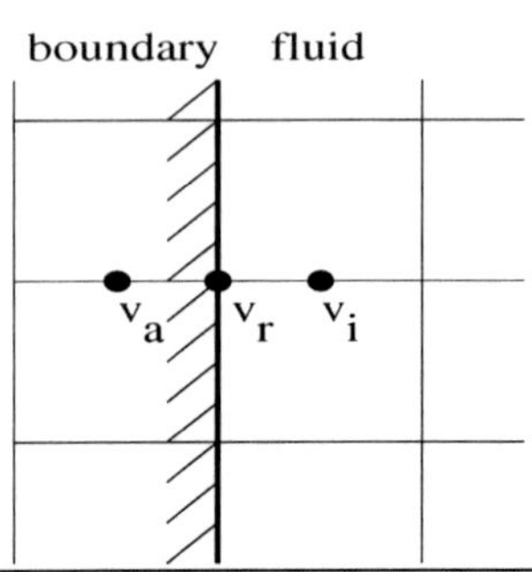

Since the vertical boundaries contain no v-values and the horizontal boundaries contain no u-values, the zero boundary value is enforced in these cases by averaging the values on either side of the boundary:

$$v_r := \frac{v_a + v_i}{2} = 0 \quad \Rightarrow \quad v_a = -v_i. \tag{3.22}$$

[20] In the case of a moving boundary, the velocities are set to nonzero values, namely, those of the wall velocities. This occurs, for example, in the driven cavity flow in Section 5.1.

On the four boundaries, we thus obtain the conditions

$$\begin{aligned} v_{0,j} &= -v_{1,j}, & v_{i_{\max}+1,j} &= -v_{i_{\max},j}, & j &= 1,\ldots,j_{\max}, \\ u_{i,0} &= -u_{i,1}, & u_{i,j_{\max}+1} &= -u_{i,j_{\max}}, & i &= 1,\ldots,i_{\max}. \end{aligned} \tag{3.23}$$

2. *Free-slip condition:* In case of a free-slip boundary condition, the velocity component normal to the boundary should vanish along with the normal derivative of the velocity component tangent to the boundary. In our rectangular domain discretized using the staggered grid, the values of velocities normal to the boundary lie directly on the boundary; hence, just as for the no-slip condition, we may set

$$\begin{aligned} u_{0,j} &= 0, & u_{i_{\max},j} &= 0, & j &= 1,\ldots,j_{\max}, \\ v_{i,0} &= 0, & v_{i,j_{\max}} &= 0, & i &= 1,\ldots,i_{\max}. \end{aligned} \tag{3.24}$$

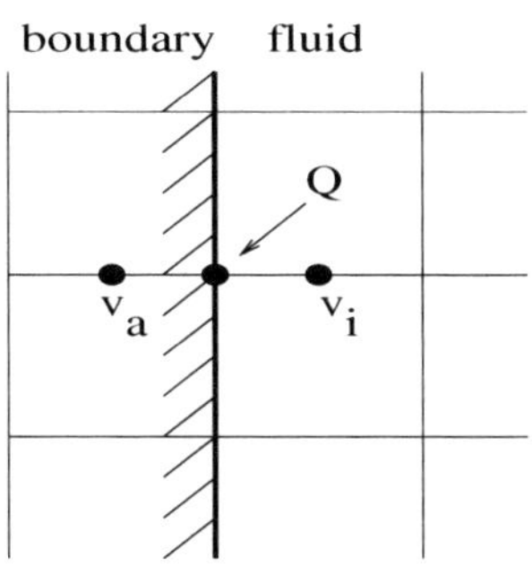

The normal derivative $\partial v/\partial n$ of the tangential velocity at a boundary point Q may be discretized by the expression $(v_i - v_a)/\delta x$; hence the requirement $\partial v/\partial n = 0$ leads to the condition

$$v_a = v_i.$$

We thus obtain the further boundary conditions

$$\begin{aligned} v_{0,j} &= v_{1,j}, & v_{i_{\max}+1,j} &= v_{i_{\max},j}, & j &= 1,\ldots,j_{\max}, \\ u_{i,0} &= u_{i,1}, & u_{i,j_{\max}+1} &= u_{i,j_{\max}}, & i &= 1,\ldots,i_{\max}. \end{aligned} \tag{3.25}$$

3. *Outflow conditions:* In the outflow boundary condition the normal derivatives of both velocity components are set to zero at the boundary, which means that the total velocity does not change in the direction normal to the boundary. In the discrete case this can be realized by setting velocity values at the boundary equal to their neighboring velocities inside the domain, i.e.,

$$\begin{aligned} u_{0,j} &= u_{1,j}, & u_{i_{\max},j} &= u_{i_{\max}-1,j}, & \\ v_{0,j} &= v_{1,j}, & v_{i_{\max}+1,j} &= v_{i_{\max},j}, & j &= 1,\ldots,j_{\max}, \\ u_{i,0} &= u_{i,1}, & u_{i,j_{\max}+1} &= u_{i,j_{\max}}, & \\ v_{i,0} &= v_{i,1}, & v_{i,j_{\max}} &= v_{i,j_{\max}-1}, & i &= 1,\ldots,i_{\max}. \end{aligned} \tag{3.26}$$

4. *Inflow conditions:* On an inflow boundary the velocities are explicitly given; we impose this for the velocities normal to the boundary (e.g., u on the left boundary) by directly fixing the values on the boundary line. For the velocity components tangential to the boundary (e.g., v at the left boundary), we achieve this by averaging the values on either side of the boundary similarly as in (3.22).

5. *Periodic boundary conditions:* For periodic boundary conditions in the x-direction, in which the boundary values on the left and right boundaries coincide, the values for u, v, and p are to be set as follows:

$$\begin{array}{lll} u_{0,j} = u_{i_{\max}-1,j}, & u_{i_{\max},j} = u_{1,j}, & p_{1,j} = p_{i_{\max},j}. \\ v_{0,j} = v_{i_{\max}-1,j}, & v_{1,j} = v_{i_{\max},j}, & v_{i_{\max}+1,j} = v_{2,j}. \end{array} \tag{3.27}$$

In contrast to the continuous problem, in which the same values are attained at $x = 0$ and $x = a$, the domain boundaries overlap here by one cell width so that the width of the domain must be chosen to exceed the period length by δx. Periodic boundary conditions in the y-direction are treated analogously.

Discretization of the Time Derivatives

We are left with the discretization of the time derivatives $\partial u/\partial t$ and $\partial v/\partial t$, for which we subdivide the time interval $[0, t_{\text{end}}]$ into equal subintervals $[n\,\delta t, (n+1)\,\delta t]$, $n = 0, \ldots, t_{\text{end}}/\delta t - 1$. This means that values of u, v, and p are considered only at times $n\,\delta t$. To discretize the time derivatives at time t_{n+1} we use *Euler's method*, which employs first-order difference quotients

$$\left[\frac{\partial u}{\partial t}\right]^{(n+1)} := \frac{u^{(n+1)} - u^{(n)}}{\delta t}, \qquad \left[\frac{\partial v}{\partial t}\right]^{(n+1)} := \frac{v^{(n+1)} - v^{(n)}}{\delta t}, \tag{3.28}$$

where the superscript (n) denotes the time level. If all remaining terms in the differential equation, in particular the spatial derivatives, are evaluated at time t_n, one obtains an *explicit method*, in which the solution values at time t_{n+1} can be computed directly from those at time t_n. In contrast, *implicit methods*, in which spatial derivatives are evaluated at time t_{n+1} instead, permit much larger time steps to be taken while still maintaining stability. In every time step, however, these methods require the solution of a linear or even nonlinear system of equations.

3.2 The Algorithm

3.2.1 The Time-Stepping Loop

Beginning at time $t = 0$ with given initial values for u and v, time is incremented by δt in each step of an outer loop until the final time t_{end} is reached. At time step n the values of all variables are known and those at time t_{n+1} are to be computed.

We begin our description of the time-stepping loop by first performing the time discretization (3.28) of the terms $\partial u/\partial t$ and $\partial v/\partial t$ in the momentum equations (2.2a,b):

$$\begin{aligned} u^{(n+1)} &= u^{(n)} + \delta t \left[\frac{1}{Re}\left(\frac{\partial^2 u}{\partial x^2} + \frac{\partial^2 u}{\partial y^2}\right) - \frac{\partial (u^2)}{\partial x} - \frac{\partial (uv)}{\partial y} + g_x - \frac{\partial p}{\partial x}\right], \\ v^{(n+1)} &= v^{(n)} + \delta t \left[\frac{1}{Re}\left(\frac{\partial^2 v}{\partial x^2} + \frac{\partial^2 v}{\partial y^2}\right) - \frac{\partial (uv)}{\partial x} - \frac{\partial (v^2)}{\partial y} + g_y - \frac{\partial p}{\partial y}\right], \end{aligned}$$

and, introducing the abbreviations

$$\begin{aligned} F &:= u^{(n)} + \delta t \left[\frac{1}{Re} \left(\frac{\partial^2 u}{\partial x^2} + \frac{\partial^2 u}{\partial y^2} \right) - \frac{\partial (u^2)}{\partial x} - \frac{\partial (uv)}{\partial y} + g_x \right], \\ G &:= v^{(n)} + \delta t \left[\frac{1}{Re} \left(\frac{\partial^2 v}{\partial x^2} + \frac{\partial^2 v}{\partial y^2} \right) - \frac{\partial (uv)}{\partial x} - \frac{\partial (v^2)}{\partial y} + g_y \right], \end{aligned} \tag{3.29}$$

we obtain the form

$$\begin{aligned} u^{(n+1)} &= F - \delta t \frac{\partial p}{\partial x}, \\ v^{(n+1)} &= G - \delta t \frac{\partial p}{\partial y}. \end{aligned} \tag{3.30}$$

To complete the discretization of the momentum equations in time, the terms on the right-hand side of (3.30) must also be associated with a time level: we evaluate F and G at time level n—i.e., *all* velocities in (3.29) belong to time level t_n—while $\partial p/\partial x$ and $\partial p/\partial y$ are associated with time level t_{n+1}. This gives us the *time discretization of the momentum equations* (2.2a,b):

$$\begin{aligned} u^{(n+1)} &= F^{(n)} - \delta t \frac{\partial p^{(n+1)}}{\partial x}, \\ v^{(n+1)} &= G^{(n)} - \delta t \frac{\partial p^{(n+1)}}{\partial y}. \end{aligned} \tag{3.31}$$

This manner of discretization may be characterized as being *explicit* in the velocities and *implicit* in the pressure; i.e., the velocity field at time step t_{n+1} can be computed once the corresponding pressure is known.

The latter is now determined by evaluating the continuity equation (2.2c) at time t_{n+1}. We substitute the relation (3.31) for the velocity field $(u^{(n+1)}, v^{(n+1)})^T$ into the continuity equation (2.2c) and obtain

$$0 = \frac{\partial u^{(n+1)}}{\partial x} + \frac{\partial v^{(n+1)}}{\partial y} = \frac{\partial F^{(n)}}{\partial x} - \delta t \frac{\partial^2 p^{(n+1)}}{\partial x^2} + \frac{\partial G^{(n)}}{\partial y} - \delta t \frac{\partial^2 p^{(n+1)}}{\partial y^2},$$

which, after rearranging, becomes a *Poisson equation for the pressure* $p^{(n+1)}$ at time t_{n+1}:[21]

$$\frac{\partial^2 p^{(n+1)}}{\partial x^2} + \frac{\partial^2 p^{(n+1)}}{\partial y^2} = \frac{1}{\delta t} \left(\frac{\partial F^{(n)}}{\partial x} + \frac{\partial G^{(n)}}{\partial y} \right). \tag{3.32}$$

In summary, the $(n+1)$st time step consists of the following parts.

Step 1: Compute $F^{(n)}$, $G^{(n)}$ according to (3.29) from the velocities $u^{(n)}$, $v^{(n)}$.

Step 2: Solve the Poisson equation (3.32) for the pressure $p^{(n+1)}$.

Step 3: Compute the new velocity field $(u^{(n+1)}, v^{(n+1)})^T$ using (3.31) with the pressure values $p^{(n+1)}$ computed in Step 2.

[21] This equation together with (3.31) ensures a divergence-free velocity field.

Solving the pressure Poisson equation in Step 2 requires boundary values for the pressure. These result from multiplying the time-discrete momentum equations (3.31) with the exterior unit normal vector $\vec{n} := (n_1, n_2)^T$ on the boundary Γ, yielding[22]

$$
\begin{aligned}
\mathrm{grad} p^{(n+1)} \cdot \vec{n} &= \frac{\partial p^{(n+1)}}{\partial x} n_1 + \frac{\partial p^{(n+1)}}{\partial y} n_2 \\
&= -\frac{1}{\delta t}\left(\left(u^{(n+1)} - F^{(n)}\right) n_1 + \left(v^{(n+1)} - G^{(n)}\right) n_2 \right).
\end{aligned}
\tag{3.33}
$$

This approach corresponds to the *Chorin projection method* developed by Chorin [Chorin, 1968] and Temam [Temam, 1969].

It is equivalent to the SMAC-method [Amsden & Harlow, 1970a], [Amsden & Harlow, 1970b] (cf. also [Tome & McKee, 1994]), in which, using the continuity equation, only a pressure correction $\delta p^{(n)}$ is computed. This is then added to $p^{(n)}$, yielding $p^{(n+1)}$. A third alternative consists in computing pressure and velocity simultaneously by iteration (cf. [Hirt & Cook, 1972], [Hirt et al., 1975]).

3.2.2 The Discrete Momentum Equations

To obtain the fully discrete momentum equations, we have yet to discretize the spatial derivatives occuring in the time-discretized momentum equations (3.31). Making use of the formulas (3.19a) and (3.19c) we obtain

$$
u_{i,j}^{(n+1)} = F_{i,j}^{(n)} - \frac{\delta t}{\delta x}\left(p_{i+1,j}^{(n+1)} - p_{i,j}^{(n+1)}\right), \qquad i = 1, \ldots, i_{\max} - 1, \quad j = 1, \ldots, j_{\max}, \tag{3.34}
$$

$$
v_{i,j}^{(n+1)} = G_{i,j}^{(n)} - \frac{\delta t}{\delta y}\left(p_{i,j+1}^{(n+1)} - p_{i,j}^{(n+1)}\right), \qquad i = 1, \ldots, i_{\max}, \quad j = 1, \ldots, j_{\max} - 1, \tag{3.35}
$$

in which the quantities F and G from (3.29) are discretized at the right and upper edges of cell (i, j), respectively:

$$
\begin{aligned}
F_{i,j} &:= u_{i,j} \\
&+ \delta t \left(\frac{1}{Re} \left(\left[\frac{\partial^2 u}{\partial x^2}\right]_{i,j} + \left[\frac{\partial^2 u}{\partial y^2}\right]_{i,j} \right) - \left[\frac{\partial (u^2)}{\partial x}\right]_{i,j} - \left[\frac{\partial (uv)}{\partial y}\right]_{i,j} + g_x \right), \\
&\qquad i = 1, \ldots, i_{\max} - 1, \quad j = 1, \ldots, j_{\max},
\end{aligned}
\tag{3.36}
$$

[22] This results in the directional derivative of the pressure in the direction of the exterior normal vector n, $\partial p/\partial n := \lim_{\varepsilon \to 0} \left(p(\vec{x} + \varepsilon \vec{n}) - p(\vec{x})\right)/\varepsilon = \mathrm{grad} p \cdot \vec{n}$. This leads to a *Neumann boundary condition*, as opposed to a *Dirichlet boundary condition*, which would prescribe the value of the function itself at the boundary. The Poisson equation with a Neumann boundary condition, $\Delta p = f$ in Ω, $\partial p/\partial n = g$ on Γ, is solvable only if the compatibility condition $\int_\Gamma g \, ds = \int_\Omega f \, dx$ holds. In this case the solution is determined uniquely up to an additive constant.

$$G_{i,j} := v_{i,j} + \delta t \left(\frac{1}{Re} \left(\left[\frac{\partial^2 v}{\partial x^2} \right]_{i,j} + \left[\frac{\partial^2 v}{\partial y^2} \right]_{i,j} \right) - \left[\frac{\partial(uv)}{\partial x} \right]_{i,j} - \left[\frac{\partial(v^2)}{\partial y} \right]_{i,j} + g_y \right), \tag{3.37}$$

$$i = 1, \ldots, i_{\max}, \quad j = 1, \ldots, j_{\max} - 1 .$$

3.2.3 The Poisson Equation for the Pressure

The discrete quantities introduced in Section 3.2.2, along with the discretization of the Laplacian (3.13), result in the discrete Poisson equation

$$\frac{p_{i+1,j}^{(n+1)} - 2p_{i,j}^{(n+1)} + p_{i-1,j}^{(n+1)}}{(\delta x)^2} + \frac{p_{i,j+1}^{(n+1)} - 2p_{i,j}^{(n+1)} + p_{i,j-1}^{(n+1)}}{(\delta y)^2} = \frac{1}{\delta t} \left(\frac{F_{i,j}^{(n)} - F_{i-1,j}^{(n)}}{\delta x} + \frac{G_{i,j}^{(n)} - G_{i,j-1}^{(n)}}{\delta y} \right), \tag{3.38}$$

$$i = 1, \ldots, i_{\max}, \quad j = 1, \ldots, j_{\max} .$$

For $i \in \{1, i_{\max}\}$ or $j \in \{1, j_{\max}\}$, the following pressure boundary values are required in (3.38):

$$\begin{array}{lll} p_{0,j}, & p_{i_{\max}+1,j}, & j = 1, \ldots, j_{\max}, \\ p_{i,0}, & p_{i,j_{\max}+1}, & i = 1, \ldots, i_{\max} . \end{array}$$

In addition, we need the following values of F and G at the boundary to compute the right-hand side of (3.38):

$$\begin{array}{lll} F_{0,j}, & F_{i_{\max},j}, & j = 1, \ldots, j_{\max}, \\ G_{i,0}, & G_{i,j_{\max}}, & i = 1, \ldots, i_{\max}, \end{array}$$

which have not yet been specified.

To determine these boundary values we look at the continuous pressure boundary condition (3.33). The resulting discretization along the left boundary (i.e., $\vec{n} = (-1, 0)^T$), for instance, is given by

$$\frac{p_{0,j}^{(n+1)} - p_{1,j}^{(n+1)}}{\delta x} = \frac{1}{\delta t} \left(u_{0,j}^{(n+1)} - F_{0,j}^{(n)} \right). \tag{3.39}$$

If we now insert this into the discrete pressure equation (3.38) for $i = 1$, we obtain

$$\frac{p_{2,j}^{(n+1)} - p_{1,j}^{(n+1)}}{(\delta x)^2} + \frac{p_{1,j+1}^{(n+1)} - 2p_{1,j}^{(n+1)} + p_{1,j-1}^{(n+1)}}{(\delta y)^2} = \frac{1}{\delta t} \left(\frac{F_{1,j}^{(n)} - u_{0,j}^{(n+1)}}{\delta x} + \frac{G_{1,j}^{(n)} - G_{1,j-1}^{(n)}}{\delta y} \right). \tag{3.40}$$

This reveals that this equation does not depend on the value of $F_{0,j}^{(n)}$, since $F_{0,j}^{(n)}$ occurs simultaneously in the right-hand side and the boundary condition of the discrete pressure equation. Hence $F_{0,j}^{(n)}$ can be selected arbitrarily. The simplest choice is $F_{0,j}^{(n)} = u_{0,j}^{(n+1)}$, which leads to $p_{0,j}^{(n+1)} = p_{1,j}^{(n+1)}$.[23] [24] The right, upper, and lower boundaries are treated analogously, which results in the following boundary values for p, F, and G:

$$\begin{array}{lll} p_{0,j} = p_{1,j}, & p_{i_{\max}+1,j} = p_{i_{\max},j}, & j = 1,\dots,j_{\max}, \\ p_{i,0} = p_{i,1}, & p_{i,j_{\max}+1} = p_{i,j_{\max}}, & i = 1,\dots,i_{\max}, \end{array} \tag{3.41}$$

and

$$\begin{array}{lll} F_{0,j} = u_{0,j}, & F_{i_{\max},j} = u_{i_{\max},j}, & j = 1,\dots,j_{\max}, \\ G_{i,0} = v_{i,0}, & G_{i,j_{\max}} = v_{i,j_{\max}}, & i = 1,\dots,i_{\max}. \end{array} \tag{3.42}$$

Due to (3.41) the pressure equation (3.38) must be modified in cells adjacent to the boundary of Ω. Using the definition of F and G at the boundary according to (3.42), we obtain

$$\begin{aligned} &\frac{\epsilon_i^E (p_{i+1,j}^{(n+1)} - p_{i,j}^{(n+1)}) - \epsilon_i^W (p_{i,j}^{(n+1)} - p_{i-1,j}^{(n+1)})}{(\delta x)^2} \\ &\qquad + \frac{\epsilon_j^N (p_{i,j+1}^{(n+1)} - p_{i,j}^{(n+1)}) - \epsilon_j^S (p_{i,j}^{(n+1)} - p_{i,j-1}^{(n+1)})}{(\delta y)^2} \\ &\qquad\qquad = \frac{1}{\delta t}\left(\frac{F_{i,j}^{(n)} - F_{i-1,j}^{(n)}}{\delta x} + \frac{G_{i,j}^{(n)} - G_{i,j-1}^{(n)}}{\delta y}\right), \\ &\qquad\qquad\qquad i = 1,\dots,i_{\max}, \quad j = 1,\dots,j_{\max}, \end{aligned} \tag{3.43}$$

in place of (3.38). The parameters

$$\epsilon_i^W := \begin{cases} 0, & i = 1, \\ 1, & i > 1, \end{cases} \quad \epsilon_i^E := \begin{cases} 1, & i < i_{\max}, \\ 0, & i = i_{\max}, \end{cases} \quad \epsilon_j^S := \begin{cases} 0, & j = 1, \\ 1, & j > 1, \end{cases} \quad \epsilon_j^N := \begin{cases} 1, & j < j_{\max}, \\ 0, & j = j_{\max}, \end{cases}$$

indicate whether cell (i, j) lies adjacent to the domain boundary, in which case the corresponding pressure values in (3.38) must be eliminated. The superscripts W, E, N, and S indicate in which direction (west, east, north, or south) the boundary lies.[25]

---

[23] The condition that the normal derivative of the pressure be zero is purely a mathematical artifact.

[24] Alternatively, $F_{0,j}^{(n)}$ could also be determined from the discrete momentum equation (3.34) requiring the value $u_{-1,j}$, which in turn can be determined from the discrete continuity equation in cell $(0, j)$. This particular boundary condition is employed in the *MAC method* [Harlow & Welch, 1965].

[25] An efficient implementation of an iterative solver for (3.43) should avoid making this distinction at every step of the iteration. All cells not adjacent to a boundary cell can, for instance, always use (3.38).

As a result, (3.43) represents a linear system of equations containing $i_{\max}j_{\max}$ equations and $i_{\max}j_{\max}$ unknowns $p_{i,j}$, $i = 1, \ldots, i_{\max}$, $j = 1, \ldots, j_{\max}$, to be solved using a suitable algorithm. For the solution of these very large, sparse linear systems of equations arising from the discretization of partial differential equations, direct methods such as Gaussian elimination are too costly in terms of computer time and storage, and thus iterative solution methods are generally applied.[26] A classical representative of these is the Gauss–Seidel method in which, starting from an initial approximation, each cell (i,j) is successively processed once in every cycle by modifying its pressure value in such a way that the corresponding equation is satisfied exactly.

An improved variant is given by the successive overrelaxation (SOR) method:

$$
\begin{aligned}
&it = 1, \ldots, it_{\max}, \\
&\quad i = 1, \ldots, i_{\max}, \\
&\qquad j = 1, \ldots, j_{\max}, \\
&\qquad\quad p_{i,j}^{\text{it}+1} := (1-\omega)\, p_{i,j}^{\text{it}} + \frac{\omega}{\left(\frac{\epsilon_i^E + \epsilon_i^W}{(\delta x)^2} + \frac{\epsilon_j^N + \epsilon_j^S}{(\delta y)^2}\right)} \\
&\qquad\qquad \cdot \left(\frac{\epsilon_i^E p_{i+1,j}^{\text{it}} + \epsilon_i^W p_{i-1,j}^{\text{it}+1}}{(\delta x)^2} + \frac{\epsilon_j^N p_{i,j+1}^{\text{it}} + \epsilon_j^S p_{i,j-1}^{\text{it}+1}}{(\delta y)^2} - rhs_{i,j} \right).
\end{aligned}
\tag{3.44}
$$

The abbreviation $rhs_{i,j}$ stands for the right-hand side of the pressure equation (3.38) in cell (i,j). The parameter ω must be chosen from the interval $[0,2]$. This choice can strongly affect the rate of convergence. The optimal choice of the relaxation parameter is discussed in Chapter 8.3 of [Stoer & Bulirsch, 1980]. The optimal choice of ω depends on the matrix of the linear system and requires eigenvalue estimates for the iteration matrix. A value often used in practice is $\omega = 1.7$. Setting $\omega = 1$ yields the Gauss–Seidel method. The iteration is terminated either once a maximal number of steps $it_{\max}$ has been taken or when the norm of the residual

$$
\begin{aligned}
r_{i,j}^{\text{it}} := {} & \frac{\epsilon_i^E (p_{i+1,j}^{\text{it}} - p_{i,j}^{\text{it}}) - \epsilon_i^W (p_{i,j}^{\text{it}} - p_{i-1,j}^{\text{it}})}{(\delta x)^2} \\
& + \frac{\epsilon_j^N (p_{i,j+1}^{\text{it}} - p_{i,j}^{\text{it}}) - \epsilon_j^S (p_{i,j}^{\text{it}} - p_{i,j-1}^{\text{it}})}{(\delta y)^2} - rhs_{i,j}\,, \\
& \qquad i = 1, \ldots, i_{\max}, \quad j = 1, \ldots, j_{\max},
\end{aligned}
\tag{3.45}
$$

has fallen below an absolute tolerance *eps* or a relative tolerance $eps\,\|p^0\|$.[27] Two commonly used norms are the discrete L^2-norm

---

[26] A thorough description of iterative methods for solving linear systems of equations can be found in the classical treatise [Varga, 1962]; for more recent results, cf., e.g., [Hackbusch, 1994] or [Barrett et al., 1993] and the references therein.

[27] For very large pressure values the absolute tolerance is too strict while for very small values it is too lenient. Including the magnitude of the pressure in the relative tolerance is thus to be recommended.

$$\|r^{\text{it}}\|_2 := \left(\frac{1}{i_{\max} j_{\max}} \sum_{i=1}^{i_{\max}} \sum_{j=1}^{j_{\max}} (r_{i,j}^{\text{it}})^2 \right)^{1/2} \tag{3.46}$$

and the maximum norm

$$\|r^{\text{it}}\|_\infty := \max \left\{ |r_{i,j}^{\text{it}}| \;\middle|\; i = 1, \ldots, i_{\max},\ \ j = 1, \ldots, j_{\max} \right\}. \tag{3.47}$$

We have used the L^2-norm together with an absolute tolerance in our simulations. Initial values for the iteration to compute the pressure values $p^{(n+1)}$ are provided by the pressure values at time level n so that, in order to compute $p^{(1)}$, we require initial values $p^{(0)}$ for the pressure (usually $p^{(0)} = 0$) in addition to those for the velocities $u^{(0)}$ and $v^{(0)}$.

This solution approach, however, introduces a difficulty: the system matrix of the linear system (3.43) is singular since the underlying boundary value problem for the pressure has only Neumann boundary conditions. This means that, for the system to have a solution, the right-hand side must lie in the range space of the matrix. In this case the solution still possesses one degree of freedom corresponding to the additive constant up to which the solution of the continuous Navier–Stokes equations is determined. If the velocity field at time t_n fails to approximately satisfy the discrete continuity equation, then the linear system has no solution and nonphysical pressure values result in the iteration (3.44).[28] Numerical experiments have shown that this difficulty can be overcome not by modifying the Poisson equation at the boundary as in (3.43), but instead by satisfying the boundary condition (3.41) for the pressure by copying the pressure values along the boundary to their neighboring cells in the boundary strip prior to each iteration step; i.e.,

$$\begin{aligned} p_{0,j}^{\text{it}+1} &= p_{1,j}^{\text{it}}, \quad p_{i_{\max}+1,j}^{\text{it}+1} = p_{i_{\max},j}^{\text{it}}, \quad j = 1, \ldots, j_{\max}, \\ p_{i,0}^{\text{it}+1} &= p_{i,1}^{\text{it}}, \quad p_{i,j_{\max}+1}^{\text{it}+1} = p_{i,j_{\max}}^{\text{it}}, \quad i = 1, \ldots, i_{\max}. \end{aligned} \tag{3.48}$$

The SOR iteration (3.44) and the calculation of the residual (3.45) then proceed as described above except that the distinction of the various cases involving the ϵ-parameters is no longer necessary, as these are now identically equal to 1.

Once the pressure values at time t_{n+1} have been calculated using either of these two methods, the velocity values u and v can be calculated according to (3.34) and (3.35).

When traditional iterative methods for solving linear systems of equations such as the Gauss–Seidel method or the SOR method are applied to systems originating from the discretization of partial differential equations, the convergence behavior deteriorates as the spacing between the grid lines is decreased. The result is that increasingly fine discretizations require more and more iteration steps to reduce the iteration error below a given tolerance.

[28]Particularly for complicated geometries such as those described in Section 3.4, it is practically impossible to find initial velocity values which satisfy the discrete continuity equation.

A new class of iterative methods for solving discrete elliptic equations was developed in the late 1970s. In these *multigrid methods*, as they are known, the number of iteration steps is independent of the number of unknowns. Thus, in general, improving the approximation of the continuous problem with a finer discretization no longer leads to an increase in the number of iteration steps. The basic idea underlying multigrid methods is that correction terms for the discrete solution on the original grid are computed on successively coarse grids obtained by doubling the spacing between the grid lines of each previous grid (cf. [Brandt, 1984], [Hackbusch, 1985], [Wesseling, 1992]).

3.2.4 The Stability Condition

In order to ensure stability of the numerical algorithm and avoid generating oscillations, stability conditions must be imposed on the stepsizes δx, δy, and δt.

In the literature one most often finds the three conditions

$$\frac{2\,\delta t}{Re} < \left(\frac{1}{\delta x^2} + \frac{1}{\delta y^2}\right)^{-1}, \qquad |u_{\max}|\,\delta t < \delta x, \qquad |v_{\max}|\,\delta t < \delta y \tag{3.49}$$

(cf., e.g., [Tome & McKee, 1994]). Here $|u_{\max}|$ and $|v_{\max}|$ are the maximal absolute values of the velocities occurring on the grid. The latter two are the famous *Courant–Friedrichs–Lewy (CFL) conditions.* They state that no fluid particle may travel a distance greater than the mesh spacing δx or δy in time δt. A detailed stability analysis can be found, e.g., in [Peyret & Taylor, 1983] or [Roache, 1976].

An adaptive stepsize control based on these stability conditions is used in [Tome & McKee, 1994]. This is implemented by selecting δt for the next time step so that each of the three conditions (3.49) is satisfied:

$$\delta t := \tau \min\left(\frac{Re}{2}\left(\frac{1}{\delta x^2} + \frac{1}{\delta y^2}\right)^{-1}, \frac{\delta x}{|u_{\max}|}, \frac{\delta y}{|v_{\max}|}\right). \tag{3.50}$$

The factor $\tau \in]0, 1]$ is a safety factor. The algorithm can, of course, also be run using a fixed given time step length which satisfies the stability condition for all time steps.

3.2.5 Summary

The entire procedure is outlined once more for clarity in Algorithm 1.

3.3 Implementation

We now turn to the implementation in the C programming language of the algorithm described above. For now, we adopt the restriction that only one type of boundary condition—no-slip, free-slip, inflow, outflow, or periodic—holds along each of the four sides of the computational domain $\Omega = [0, a] \times [0, b]$.

```
Set t := 0, n := 0
Assign initial values to u, v, p
While t < t_end
    Select δt (according to (3.50) if stepsize control is used)
    Set boundary values for u and v
    Compute F^(n) and G^(n) according to (3.36),(3.37)
    Compute the right-hand side of the pressure equation (3.38)
    Set it := 0
    While it < it_max and ||r^it|| > eps (resp., ||r^it|| > eps ||p^0||)
        Perform an SOR cycle according to (3.44)
        Compute the residual norm for the pressure equation ||r^it||
        it := it + 1
    Compute u^(n+1) and v^(n+1) according to (3.34),(3.35)
    t := t + δt
    n := n + 1
```

Algorithm 1. Base version.

3.3.1 Problem Parameters and Data Structures

The algorithm requires the definition of the following quantities, in which `REAL` defines a floating point number by way of the *macro definition* `#define REAL float` or `double`, respectively.

- Geometry data:

`REAL`	`xlength`	domain size in x-direction,
`REAL`	`ylength`	domain size in y-direction,
`int`	`imax`	number of interior cells in x-direction,
`int`	`jmax`	number of interior cells in y-direction,
`REAL`	`delx`	length δx of one cell in x-direction,
`REAL`	`dely`	length δy of one cell in y-direction.

- Time-stepping data:

`REAL`	`t`	current time value,
`REAL`	`t_end`	final time t_{end},
`REAL`	`delt`	time step size δt,
`REAL`	`tau`	safety factor for time step size control τ.

- Pressure-iteration data:

`int`	`itermax`	maximal number of pressure iterations in one time step,
`int`	`it`	SOR iteration counter,
`REAL`	`res`	norm of pressure equation residual,
`REAL`	`eps`	stopping tolerance *eps* for pressure iteration,
`REAL`	`omg`	relaxation parameter ω for SOR iteration,
`REAL`	`gamma`	upwind differencing factor γ.

- Problem-dependent quantities:

`REAL`	`Re`	Reynolds number *Re*;
`REAL`	`GX, GY`	body forces g_x, g_y, (e.g., gravity);
`REAL`	`UI, VI, PI`	initial data for velocities and pressure;
`int`	`wW, wE, wN, wS`	specify the type of boundary condition along the western (left), eastern (right), northern (upper), and southern (lower) boundaries of Ω = [0,`xlength`] $\times$ [0,`ylength`]; each may have one of the values: 1 for free-slip conditions, 2 for no-slip conditions, 3 for outflow conditions, 4 for periodic boundary conditions;
`char`	`problem`	This variable allows further flow-specific quantities, such as inflow velocity or internal obstacles, to be specified depending on the problem type.

For inflow boundary conditions, the velocities must be set to fixed given values (e.g., a parabolic or constant inflow profile), which may depend on the specific problem being solved; hence the setting of the velocity values at the boundary must take place in a problem-specific function and cannot be prescribed completely in terms of the flags `wW,wE,wN,` and `wS`.

Except for the quantities `t`, `it`, `delx`, and `dely`, these parameters are all supplied in an input file `inputfile` which can be made available to the program as a command-line argument. Furthermore, the following arrays of dimension `[0,imax+1]x[0,jmax+1]` will be used as data structures.

- Data arrays:

`REAL`	`**U`	velocity in x-direction,
`REAL`	`**V`	velocity in y-direction,
`REAL`	`**P`	pressure,
`REAL`	`**RHS`	right-hand side for pressure iteration,
`REAL`	`**F,**G`	F, G.

Memory for these variables should be *allocated dynamically*; i.e., no fixed maximal memory size is reserved at compile time. Rather, only as much memory as is needed is allocated at program start. We indicate one possible way for doing this, which can be found in [Press et al., 1990].

```
REAL **RMATRIX(int nrl,int nrh,int ncl,int nch){
/** reserves memory for matrix of size [nrl,nrh]x[ncl,nch] **/
  int i;
  REAL **m;
  /**** allocate row pointers ****/
  if((m=(REAL**)
         malloc((unsigned)(nrh-nrl+1)*sizeof(REAL*)))==NULL){
    printf("no more memory \n");
    exit(0);}
  m -= nrl;
  /** allocate rows and set previously allocated row pointers
                                            to point to these **/
  for(i=nrl;i<=nrh;i++){
    if((m[i]=(REAL*)
          malloc((unsigned)(nch-ncl+1)*sizeof(REAL)))==NULL){
      printf("no more memory\n");
      exit(0);}
    m[i] -= ncl;}
  /** return pointer to the array of row pointers **/
  return m;
}
```

Clean memory management mandates that memory be freed when it is no longer needed. The function for freeing memory allocated by `RMATRIX` reads as follows:

```
void FREE_RMATRIX(REAL **m, int nrl,int nrh,int ncl,int nch){
/** frees memory of matrix allocated by RMATRIX **/
  int i;
  for(i=nrh; i>nrl; i-) free((char*) (m[i]+ncl));
  free((char*) (m+nrl));
}
```

Typical calls to these functions would be "`U = RMATRIX(0,imax,0,jmax);`" and "`FREE_RMATRIX(U,0,imax,0,jmax);`".

3.3.2 The Program

Using the variables and data structures defined above, the following functions are to be written.[29]

1. `int READ_PARAMETER(char *inputfile,char *problem,xlength,ylength, imax,jmax,delx,dely,delt,t_end,tau,itermax,eps,omg,gamma,Re,GX,GY, UI,VI,PI,wW,wE,wN,wS)`: The listed quantities are read from the input file passed by the variable `inputfile`, and `delx` and `dely` are determined from

[29]For brevity we have omitted the type specification in the argument list for variables previously introduced. To avoid such lengthy argument lists, related variables can be grouped using the `struct` language feature.

`imax` and `xlength` or `jmax` and `ylength`, respectively. The return value may be used for error handling.

2. `void INIT_UVP(U,V,P,imax,jmax,UI,VI,PI)`: The arrays `U,V,P` are initialized to the constant values `UI,VI,` and `PI` on the entire domain.

3. `void COMP_DELT(delt,imax,jmax,delx,dely,U,V,Re,tau)`: The stepsize `delt` for the next time step is calculated according to (3.50). In case of negative `tau` the stepsize read in `READ_PARAMETER` is to be used.

4. `void SETBCOND(U,V,imax,jmax,wW,wE,wN,wS)`: The boundary values for the arrays `U` and `V` are set depending on the boundary data parameters `wW,wE,wN,wS` according to the formulas in Section 3.1.2.

5. `void SETSPECBCOND(U,V,imax,jmax,char *problem)`: Here, specific boundary conditions (such as inflow boundary conditions) of the problem specified by the parameter `problem` can be set. It is also possible to redefine the boundary data on parts of a boundary which has been previously assigned a boundary condition by one of the flags `wW`, `wE`, `wN`, or `wS` (for example, a small in- or outflow area within a solid wall).

6. `void COMP_FG(U,V,F,G,imax,jmax,delt,delx,dely,GX,GY,gamma,Re)`: Computation of `F` and `G` according to (3.36) and (3.37). At the boundary the formulas (3.42) must be applied.

7. `void COMP_RHS(F,G,RHS,imax,jmax,delt,delx,dely)`: Computation of the right-hand side of the pressure equation (3.38).

8. `int POISSON(P,RHS,imax,jmax,delx,dely,eps,itermax,omg,res)`: SOR iteration for the pressure Poisson equation according to (3.44). The iteration is terminated once the residual norm `res` drops below the tolerance limit `eps` (absolute or relative, multiplied by the norm of the initial pressure) or once the maximal number of iterations `itermax` is reached. Upon completion, the number of steps taken is returned and the current residual norm is stored in `res`.
   If the pressure boundary values are treated using the second method, then the boundary values must be set according to (3.48) prior to each iteration step.

9. `void ADAP_UV(U,V,F,G,P,imax,jmax,delt,delx,dely)`: The new velocities are calculated according to (3.34) and (3.35).

Finally, these functions are combined to implement the algorithm given above in the main program `main`.

3.3.3 Guidelines for Modular Programming

In order to keep the evolving program flexible and manageable and to avoid long compilation times during development, it is useful to collect the functions in separate modules (files), which are then linked with the main program after compilation. As an example, the functions READ_PARAMETER, RMATRIX, and INIT_UVP could be grouped into a file init.c. Once this module has been written and compiled, it then no longer needs to be recompiled each time a detail in another part of the program is modified.

The following modular structure suggests itself for the program to be developed here (also with regard to extensions to be added later):

init.c:	READ_PARAMETER, RMATRIX, FREE_RMATRIX, INIT_UVP,
boundary.c:	SETBCOND and SETSPECBCOND,
uvp.c:	COMP_FG, COMP_RHS, POISSON, and ADAP_UV,
main.c:	main program.

If a function defined in a module B.c is needed in another module A.c, then the corresponding function declaration must be made known to module A.c. This is usually accomplished with the help of so-called *header files*: all function declarations in a module A.c which are needed in other modules are collected in a file A.h which is then included in all calling modules using the (preprocessor) directive #include "A.h". This saves having to explicitly list all required function declarations in each file.

In a UNIX environment the compilation and linking processes can be automated with the help of a *makefile*, which is stored in a file of the same name. For our purposes the makefile could look as follows:

```
CC = gcc
CFLAGS = -Wall -pedantic
.SUFFIXES: .o .c
.c.o:  ; $(CC) -c $(CFLAGS) $*.c
OBJ = init.o boundary.o uvp.o main.o

run: $(OBJ)
        $(CC) $(CFLAGS) -o run $(OBJ) -lm

init.o     : init.h
boundary.o: boundary.h
uvp.o      : uvp.h
main.o     : init.h boundary.h uvp.h
```

Explanations of the individual commands:

- CC = gcc and CFLAGS = -Wall -pedantic are *macro definitions* selecting a particular compiler (in this case, the GNU C compiler) and various possible compilation options. In this case, the options -Wall and -pedantic cause the compiler to generate warnings whenever it encounters possible sources of error or possibly incorrect code. The option -O causes various optimizations with the goal of minimizing the runtime to be performed.

- `.SUFFIXES: .o .c` determines a general pattern of dependencies among files. Whenever a .c file is modified, the corresponding .o (object code) file with the same name must be rebuilt.
- `.c.o: ; $(CC) -c $(CFLAGS) $*.c` causes the *transformation* of .c files to .o files by means of the command `$(CC) -c $(CFLAGS) $*.c`. Whichever C compiler is defined in the macro `$(CC)` is inserted in place of the macro variable `$(CC)`.
- `OBJ = init.o boundary.o uvp.o main.o` is another macro definition.
- `run: $(OBJ)` determines the dependencies of the program `run`. Whenever one of the .o files in `OBJ` is modified, the program must be linked again.
- `$(CC) $(CFLAGS) -o run $(OBJ) -lm` is the command to link the object files listed in the macro `OBJ` together. The option `-o run` causes the executable program thus generated to be named `run` (the default executable name being `a.out`), and `-lm` links the mathematical library.

  IMPORTANT: This command must begin with a TAB !
- The remaining commands define special dependencies. For example, if `uvp.h` is modified, then `uvp.o` and `main.o` must both be rebuilt.

More information on this subject area can be found in the many UNIX and C books available.

3.4 Treatment of General Geometries

Up to now we have described the simulation of flows in rectangular domains Ω. A simple extension will enable us to approximately treat flows in arbitrary two-dimensional geometries.

3.4.1 Introduction of Obstacle Domains

For this we imbed the flow domain Ω in a rectangle $\mathcal{G}$ of smallest possible size, which we cover with a grid as described in Section 3.1. The cells of $\mathcal{G}$ are then divided into *fluid cells* (which lie completely or mostly in Ω) and *obstacle cells* (which lie completely or mostly in the obstacle $\mathcal{H} := \mathcal{G} \setminus \Omega$). The Navier–Stokes equations are then solved only in the fluid cells. The cells $(i,j), i \in \{0, i_{\max}+1\}$ or $j \in \{0, j_{\max}+1\}$, belonging to the artificial boundary strip are also considered obstacle cells.

Furthermore, we denote those obstacle cells which share an edge (*boundary edge*) with at least one fluid cell as *boundary cells.* A domain Ω possessing an arbitrary curved boundary is thus approximated by a domain $\tilde{\Omega}$ whose boundary is specified by the set of boundary edges lying on grid lines (see Figure 3.7).[30]

In fluid cells adjoining obstacle cells we require the following boundary values in order to compute F and G according to (3.36), (3.37):

[30] A more accurate resolution of curved boundaries is described, e.g., in [Tome & McKee, 1994].

- values of the normal velocity components at the boundary edges and
- values of the tangential velocity components on edges between boundary cells.

Depending on the type of boundary condition, we use the discretizations described in Section 3.1.2. In the example depicted in Figure 3.8 boundary edges are marked with a square, and the edges between two boundary cells with a circle.

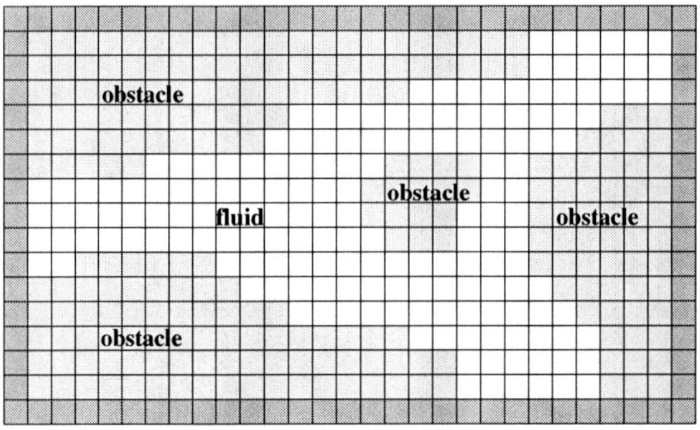

FIG. 3.7. *Approximate imbedding of an arbitrary domain into a rectangular domain.*

Moreover, the F and G values at the boundary edges are needed to compute the pressure at the centers of the fluid cells bordering on boundary cells, and for these cells the pressure equation must be modified analogously to (3.43). In this case, however, the parameters $\epsilon^N, \epsilon^S, \epsilon^W$, and ϵ^E depend on i *and* j and are set to zero whenever the corresponding neighboring cell is an obstacle cell; otherwise, they are set to one. As an alternative to modifying the pressure equation in cells close to the boundary, one can set the pressure in each boundary cell to the value of the pressure in the neighboring fluid cell.

To visualize the flow using particle tracing and streaklines that will be introduced later (see Section 4.2), we will require not only the velocity values on edges between two boundary cells but also those on edges of which one end coincides with a fluid cell corner, in order to interpolate the velocity values described in Section 4.2.1. These edges have also been marked with a circle in Figure 3.8.

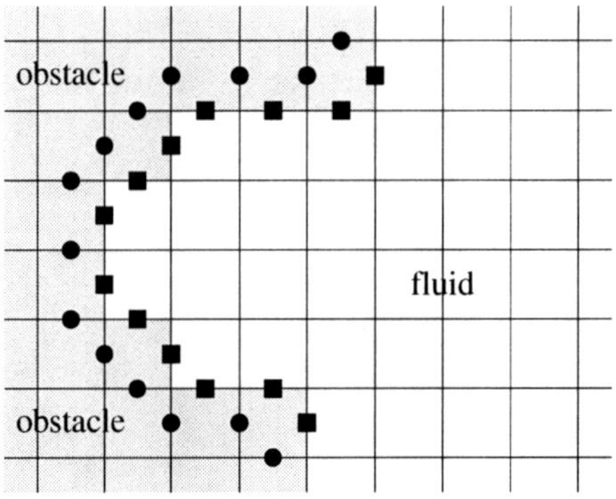

FIG. 3.8. *Required boundary values.*

To identify the fluid cells we use an integer array `int **FLAG`, which can be initialized as follows:

- `C_F` for a fluid cell and
- `C_B` for an obstacle cell.

The macros `C_B` and `C_F` denote fixed integer constants which can be chosen arbitrarily.

The set of boundary cells must now be further classified to determine the boundary values. To this end, we store information about which of the four adjacent cells are fluid cells in the flag array. The macro `B_E`, for instance, may denote a boundary cell whose right (eastern) neighbor belongs to $\tilde{\Omega}$, or `B_SW` may denote a boundary cell whose lower (southern) and left (western) neighbors belong to $\tilde{\Omega}$. We assume the discretization to be fine enough that only boundary cells with one neighboring fluid cell (*edge cells*) or two neighboring fluid cells sharing a corner (*corner cells*) can occur.[31] Boundary cells with two opposite or even three or four neighboring fluid cells are excluded (*inadmissible boundary cells*), since these would no longer lead to uniquely definable boundary values.

We illustrate the practical computation of the boundary values using the no-slip condition as an example.[32] (See Figure 3.9.) For a northern edge cell (i,j) carrying the flag `B_N`, we set, in accordance with the formulas (3.21), (3.23), and (3.42),

$$v_{i,j} = 0, \qquad u_{i-1,j} = -u_{i-1,j+1}, \qquad u_{i,j} = -u_{i,j+1}, \qquad G_{i,j} = v_{i,j}, \tag{3.51}$$

or, for a western edge cell (i,j) with flag `B_W`,

$$u_{i-1,j} = 0, \quad v_{i,j-1} = -v_{i-1,j-1}, \quad v_{i,j} = -v_{i-1,j}, \quad F_{i-1,j} = u_{i-1,j}. \tag{3.52}$$

The boundary values for `B_S`- and `B_E`-cells are set analogously.

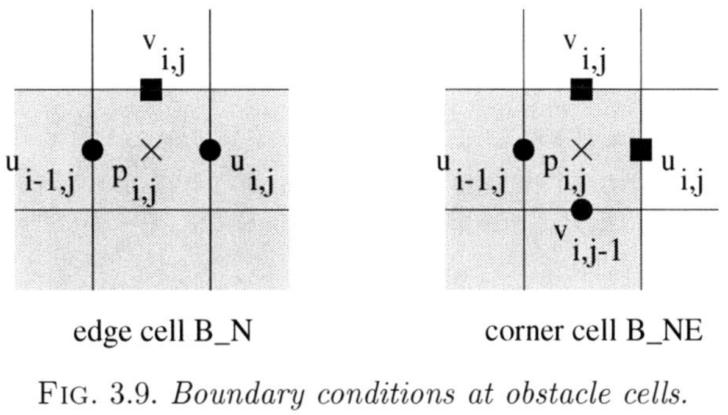

FIG. 3.9. *Boundary conditions at obstacle cells.*

For corner cells we apply the condition to the normal velocity component at the two edges adjacent to fluid cells, and the condition to the tangential velocity component at the remaining two edges.[33]

[31] We consider those cells to be *neighboring* which share a common edge, not only a corner.

[32] The treatment of other boundary conditions proceeds analogously. If several boundary conditions are to be implemented then the set of boundary cell types must be subdivided further.

[33] To avoid computing these velocities repeatedly it suffices to set the tangential velocity components only at the northern and eastern edges of corner and edge cells, respectively.

Thus, for a cell (i,j) with flag `B_NE`, for example, we obtain

$$\begin{aligned} u_{i,j} &= 0, & u_{i-1,j} &= -u_{i-1,j+1}, & F_{i,j} &= u_{i,j}, \\ v_{i,j} &= 0, & v_{i,j-1} &= -v_{i+1,j-1}, & G_{i,j} &= v_{i,j}. \end{aligned} \tag{3.53}$$

If the pressure equation is not modified in the boundary cells as in (3.43), then the pressure in the edge cells must be set equal to the pressure in the neighboring fluid cells prior to each SOR step; i.e., for example,

for flag `B_N`: $p_{i,j} = p_{i,j+1}$ and for flag `B_W`: $p_{i,j} = p_{i-1,j}$.

In corner cells we then use an average of the two conditions for the edge cells; i.e., for example,

for flag `B_NE`: $p_{i,j} = (p_{i,j+1} + p_{i+1,j})/2$.

The pressure equation (3.38) is solved in the fluid cells while the computation of the F- and G-values (3.36), (3.37) as well as the update of the velocity values (3.34), (3.35) occurs only on edges between two fluid cells.

To summarize, we collect all cell and edge definitions once more:

- fluid cell: cell lying completely or mostly in Ω,
- obstacle cell: cell lying completely or mostly in $\mathcal{G} \setminus \Omega$,
- boundary cell: obstacle cell bordering on a fluid cell,
- edge cell: boundary cell bordering on exactly one fluid cell,
- corner cell: boundary cell bordering on two fluid cells sharing a corner,
- inadmissible boundary cell: boundary cell bordering on fluid cells along opposite edges,
- boundary edge: edge between a fluid and a boundary cell.

3.4.2 Implementation

1. The integer array `int **FLAG` of dimension `[0,imax+1]x[0,jmax+1]` is now needed as an additional data structure. It is convenient here to use a binary representation for the possible states, assigning one bit to each cell and its four neighbors, e.g.,

center	east	west	south	north

   Each bit can be set to either one, when the corresponding cell is a fluid cell, or zero, when it is an obstacle cell. The cells in the interior of the obstacle would thus carry the flag 00000, boundary cells (including inadmissible ones), a flag between 00001 and 01111, fluid cells bordering on obstacle cells, a flag between 10000 and 11110, and cells in the interior of the fluid region, a flag value of 11111. The flag `B_SE`, for instance, corresponds to a bit coding of 01010 with a decimal value of 10. Using the bit manipulation operations available in the C programming language, these case distinctions can be expressed rather concisely.

2. To identify the fluid region we shall need two new functions. These can be incorporated into the file `init.c`.

   i. `int **IMATRIX(nrl,nrh,ncl,nch)`: Memory is allocated for an integer array of dimension `[nrl,nrh]x[ncl, nch]` by the same technique as used in the function `RMATRIX`. This function is called in the initialization phase of the main program.

   ii. `void INIT_FLAG(problem,FLAG,imax,jmax)`: The array `FLAG` is initialized with the flags `C_F` for fluid cells and `C_B` for obstacle cells as specified by the parameter `problem`. This can be done either by explicit specification of the problem geometry or by reading data from a file, in which the cells are divided into fluid cells and obstacle cells. This is followed by a loop over all cells in which the boundary cells are marked with the appropriate flags `B_xy` depending on in which direction neighboring fluid cells lie. This function is also called in the initialization phase of `main`.

3. In addition, the following functions must be adapted:

   i. In `SETBCOND` and `COMP_FG`, the assignment of the boundary values in the boundary cells defined by the flag array as described above must be added.

   ii. In `POISSON`, the parameters $\epsilon_{i,j}^N, \epsilon_{i,j}^S, \epsilon_{i,j}^W$, and $\epsilon_{i,j}^E$ must be determined in fluid cells adjoining obstacle cells prior to each relaxation step (3.44). If the second approach for treating the pressure boundary condition is used the boundary conditions must be set in the obstacle cells before each iteration step.

   iii. In `POISSON`, the SOR iteration and residual calculation are restricted to the fluid cells, while the F- and G-values in `COMP_FG` and the velocities in `ADAP_UV` are only calculated on edges separating two fluid cells. Note also that the normalization of the residual in `POISSON` is obtained not by dividing by $i_{\max} j_{\max}$, but by the actual number of fluid cells.

Chapter 4

Visualization Techniques

The evaluation of computed data is an essential aspect of numerical flow simulation. To handle the vast amounts of data generated in the course of a computation, one can use some simple visualization techniques that are described in Section 4.1. Beyond these, we introduce further visualization techniques that have been developed specifically for flow applications in Sections 4.2 and 4.3.

4.1 Standard Techniques

4.1.1 Real-Valued Functions

To obtain an image of a real-valued function defined on a two-dimensional domain $\Omega \subset \mathbb{R}^2$,

$$f : \Omega \to \mathbb{R}\,, \quad (x, y) \mapsto f(x, y)\,,$$

such as the pressure p or temperature T (to be introduced later), the *graph of the function*

$$G_f := \{(x, y, z) \in \Omega \times \mathbb{R} \mid z = f(x, y)\} \subset \mathbb{R}^3$$

is often used. In this case it describes a two-dimensional surface over Ω obtained by lifting each point $(x, y) \in \Omega$ to its function value $f(x, y)$ (see Figure 4.1, left). Similarly, the *level sets* belonging to $c \in \mathbb{R}$,

$$N_f(c) := \{(x, y) \in \Omega \mid f(x, y) = c\} \subset \mathbb{R}^2,$$

consisting of all points $(x, y) \in \Omega$ at which the function f takes the value c, can be used as well. Whenever f is a continuous function, the level set $N_f(c)$ can be locally represented as a curve:

$$\varphi_c : [a, b] \to N_f(c)\,, \quad s \mapsto \varphi_c(s) = \begin{pmatrix} x_c(s) \\ y_c(s) \end{pmatrix} \in N_f(c)\,. \tag{4.1}$$

Such a curve is known as a *contour line*, and the following relations hold:

$$f(\varphi_c(s)) = f(x_c(s), y_c(s)) = c \qquad \forall s \in [a, b] \tag{4.2}$$

(see Figure 4.1, middle). If each function value of f is assigned a color or shade of gray, an image analogous to the contour lines results (see Figure 4.1, right).

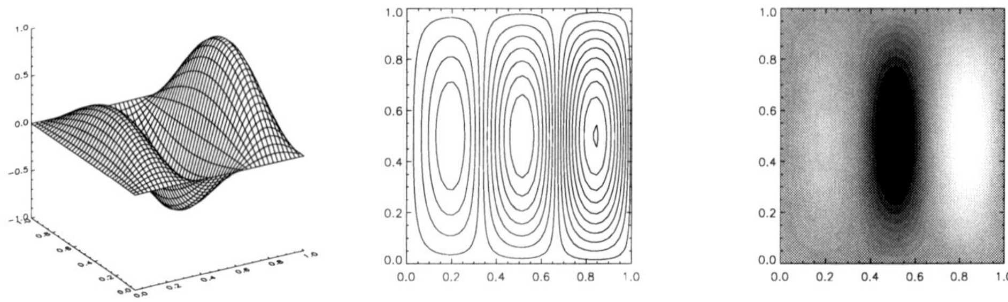

FIG. 4.1. *The graph, contour lines, and a grayscale image of the function* $f(x,y) = \sqrt{x}\sin(3\pi x)\sin(\pi y)$ *on* $\Omega = [0,1] \times [0,1]$.

For real-valued functions defined on a three-dimensional domain $\Omega \subset \mathbb{R}^3$, visualization becomes more difficult. The graph of the function is no longer suitable for representing the function, as the level sets now become two-dimensional surfaces in $\mathbb{R}^3$ known as *isosurfaces.* The calculation of these isosurfaces is usually very computation-intensive. When the three-dimensional domain is intersected with planes, the techniques for functions of two variables can be applied to the resulting planes of intersection. To avoid losing sight of the three-dimensional character of the data, it is useful to be able to view these intersecting planes from different perspectives in a three-dimensional frame or to move the planes around. Visualization techniques for the three-dimensional case can be found in Section 11.4.

4.1.2 Vector-Valued Functions

For vector-valued functions from $\Omega \subset \mathbb{R}^2$ to $\mathbb{R}^2$,

$$\vec{f} : \Omega \to \mathbb{R}^2, \quad (x,y) \mapsto \vec{f}(x,y) = \begin{pmatrix} f_1(x,y) \\ f_2(x,y) \end{pmatrix},$$

such as the velocity field $\vec{u} = \binom{u}{v}$, the methods of the previous section can be applied individually to each vector component, or, alternatively, the vector field $\vec{f}$ can be made visible by attaching vectors to selected points in the domain of definition (see Figure 4.2). These vectors can be attached to either points of a regular grid or randomly chosen points. To avoid overly long vectors, the vector field $\vec{f}$ should be suitably scaled.

For three-dimensional domains or image spaces the techniques for two-dimensional vector fields can again be used by projecting the three-dimensional image onto a two-dimensional viewing plane. Otherwise one could again use two-dimensional sections to which the previously mentioned techniques are applied.

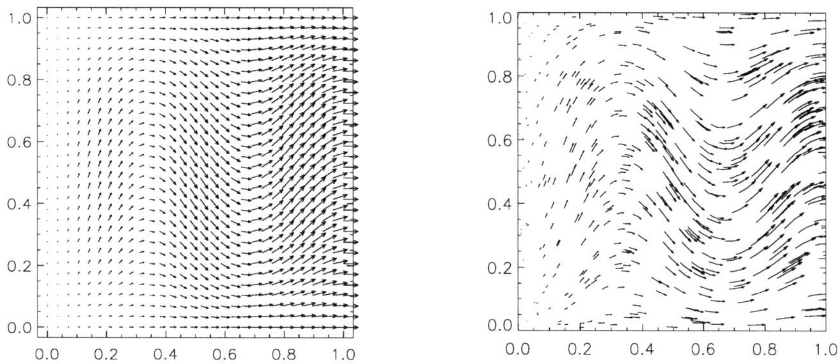

FIG. 4.2. *Two vector representations of the vector field* $\vec{g}(x, y) = \binom{x}{f(x,y)}$ *with the real-valued function* f *from Figure* 4.1: *using vectors at regularly spaced points (left) and at randomly chosen points (right).*

Moreover, the techniques for three-dimensional real-valued functions can be applied to the individual components of vector-valued functions. Examples of three-dimensional visualization can be found in Section 11.4.

4.1.3 Graphics Tools and Standards

Besides current graphics standards such as *GL* (*Graphics Library*), *OpenGL*, *Starbase*, or the earlier *GKS* (*Graphics Kernel System*) and *Phigs* systems, a variety of graphics packages have established themselves on the market. Most of these (*AVL*, *Explorer*, *IDL*, and *Grape* could be mentioned) offer advanced three-dimensional visualization techniques. Despite its strong orientation towards two dimensions, the graphics tool *gnuplot* deserves special mention since it is available free of charge for most computer platforms.

In addition, many mathematics tools possess their own graphics capabilities; examples are *Mathematica, Maple, MATLAB,* and *CLAM.* The capabilities and features of each of these tools vary quite a bit, but almost all offer the simple standard techniques introduced in the preceding sections, 4.1.1 and 4.1.2.

Flow visualization also has its specially developed and refined tools such as *VISUAL3, cfdExplorer, CFD-View, CFView, Fieldview, FlowEyes,* and *tecplot.*

But even in the absence of such tools, in most programming languages libraries can be used that support some simple (and even enhanced) form of graphics. Essentially, the programming of the two-dimensional methods of Sections 4.1.1 and 4.1.2 requires only plotting lines, setting points, and defining coordinate systems. Representing a vector field $\vec{f}(x, y) = (f_1(x, y), f_2(x, y))^T$ (see Figure 4.2) requires only drawing lines whose starting points $\vec{a}_{i,j}$ lie on the grid points (x_i, y_j) and whose end points are determined by the vectors $\vec{f}(x_i, y_j) = (f_1(x_i, y_j), f_2(x_i, y_j))^T$: $\vec{e}_{i,j} = \vec{a}_{i,j} + \vec{f}(x_i, y_j)$. A suitable scaling of the length of these vector fields must be carried out to avoid cluttering the image by overly long vectors.

Suggestions for how to compute and plot contour lines of a real-valued function can be found in [Pavlidis, 1982]. Another part of the standard graphics repertoire is *hidden line removal* algorithms required for plotting the graphs (see Figure 4.1) of real-valued functions of two variables (cf. [Foley & van Dam, 1984], [Preparata & Shamos, 1985], and [Harrington, 1983]).

4.1.4 Implementation

To apply the standard techniques introduced in Sections 4.1.1 and 4.1.2 to the results of a flow calculation, the data to be visualized must be exported. The simplest method for doing this lies in writing the data out to a file.[34] This file is then read as needed by the graphics tool. Including auxiliary data such as the size of the data arrays (`imax` and `jmax`), `xlength`, and `ylength` is often helpful for the proper scaling of the resulting image. The following function is to be coded:

> `void OUTPUTVEC(U,V,P,FLAG,xlength,ylength,imax,jmax,outputfile)`:
> The file specified in the argument `outputfile` of type `char*` should contain the values `imax`, `jmax`, `xlength`, and `ylength` as a header followed by the current values of the arrays `U`, `V`, `P`, and `FLAG`. For visualization, all these values should be those at cell centers, which requires the `U`- and `V`-values to be interpolated from their edge values with the formula $(w_1 + w_2)/2$. To easily visualize the structure of the obstacles, the array values belonging to obstacle cells can be set to special designated values (e.g., one somewhat smaller than the minimal value in the array), or the obstacle structure encoded in the flag array can be superimposed on the image if the visualization software has this capability.
>
> *Note.* To read the data written out to disk, many visualization tools will accept data in *binary format.* For large data sets, this saves considerable time in the reading and writing phases and dramatically reduces storage requirements. If different computer platforms are used for computation and visualization, however, then the data must be held in an input and output format understood by both systems; usually this will be ASCII format.

The function `OUTPUTVEC` can be placed in a new module `visual.c`.

4.2 Flow Visualization by Particle Tracing and Streaklines

When studying flows such as the flow past an obstacle described in Section 5.3, one is often interested in analyzing the flow pattern in terms of the movement of individual fluid particles. We consider two methods for doing this.

- *Pathlines (particle tracing):* The positions of *one* particle throughout successive points in time are shown.

[34] If, for the purpose of real-time visualization, the data is to be sent directly to the visualization tool, then direct main-memory connections, known as *pipes*, linking the two processes (numerical simulation and visualization) are a possible solution (cf., e.g., [Stevens, 1990]).

- *Streaklines:* Particles are injected into the flow at a fixed location in short regular time intervals, and the positions of those particles belonging to the *same injection time* are shown.[35]

The basis for the computation of such particle movements lies in calculating the position of a particle at time $t_{n+1} = t_n + \delta t$ given its position $(x^{(n)}, y^{(n)})$ at time t_n. This occurs in two steps:

1. Determine the velocities $u^{(n)}, v^{(n)}$ at position $(x^{(n)}, y^{(n)})$ in the flow field.
2. Compute the new position of this particle at time t_{n+1} using one step of the (forward) Euler method:

$$x^{(n+1)} = x^{(n)} + \delta t\, u^{(n)}, \quad y^{(n+1)} = y^{(n)} + \delta t\, v^{(n)} . \tag{4.3}$$

The second of these steps is rather simple; the first, however, warrants some explanation.

4.2.1 Interpolation of Velocities in a Staggered Grid

Since the velocities u and v are available only at certain points of the staggered grid, the velocities required for the particle position updates must somehow be approximated using the known ones. We are thus given the

- particle position (x, y) and the
- velocities $u_{i,j}$, $v_{i,j}$ on the staggered grid:

$$\begin{aligned}
u_{i,j} = u(x_i, y_j) \quad &\text{with} \quad x_i = i\,\delta x, \quad y_j = \left(j - \frac{1}{2}\right)\delta y, \\
&\qquad i = 0, \ldots, i_{\max}, \; j = 0, \ldots, j_{\max} + 1, \\
v_{i,j} = v(x_i, y_j) \quad &\text{with} \quad x_i = \left(i - \frac{1}{2}\right)\delta x, \quad y_j = j\,\delta y, \\
&\qquad i = 0, \ldots, i_{\max} + 1, \; j = 0, \ldots, j_{\max},
\end{aligned}$$

and we seek the velocities $u(x, y)$, $v(x, y)$ at position (x, y).

1. *Computation of $u(x, y)$:* The u-velocity at position (x, y) is computed by interpolating the four nearest discrete u-velocities. This is done by first determining the u-cell surrounding this location, i.e., the (smallest) rectangle containing (x, y), the corners of which carry discrete u-velocities (see Figures 4.3 and 4.4). We denote the index of the upper right corner by (i, j), which may be determined using the integer operations[36]

[35] The term "streakline" originates from the fact that one wishes to observe the evolution of streaks of dye introduced into the fluid at different instances. One way of realizing this in a physical experiment is to introduce small air bubbles into the fluid along a thin wire (which doesn't affect the flow).

[36] The notation (*int*) is borrowed from the C programming language. In the case considered here, this *cast operator* is applied exclusively to positive real numbers so that it coincides with the *Gauss bracket* $[x]$: $[x] := \max\{z \in \mathbb{Z} \mid z \leq x\}$ for $x \in \mathbb{R}$.

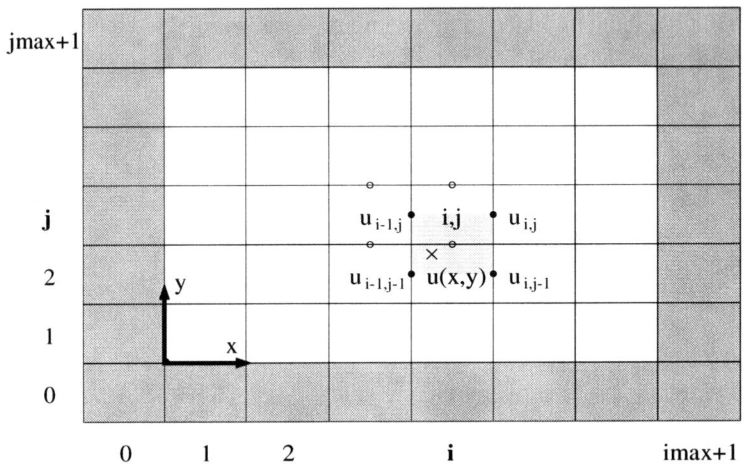

FIG. 4.3. *A u-cell (i,j) (gray shading) and corners of a v-cell (circles) in the staggered grid.*

$$i := (int)\left(\frac{x}{\delta x}\right) + 1, \quad j := (int)\left(\frac{y + \frac{\delta y}{2}}{\delta y}\right) + 1.$$

The coordinates of the cell corners are thus

$$x_1 := (i-1)\,\delta x, \qquad y_1 := ((j-1) - \tfrac{1}{2})\,\delta y,$$
$$x_2 := i\,\delta x, \qquad y_2 := (j - \tfrac{1}{2})\,\delta y,$$

on which the four *staggered grid velocities*

$$u_1 := u_{i-1,j-1}, \quad u_2 := u_{i,j-1},$$
$$u_3 := u_{i-1,j}, \quad u_4 := u_{i,j}$$

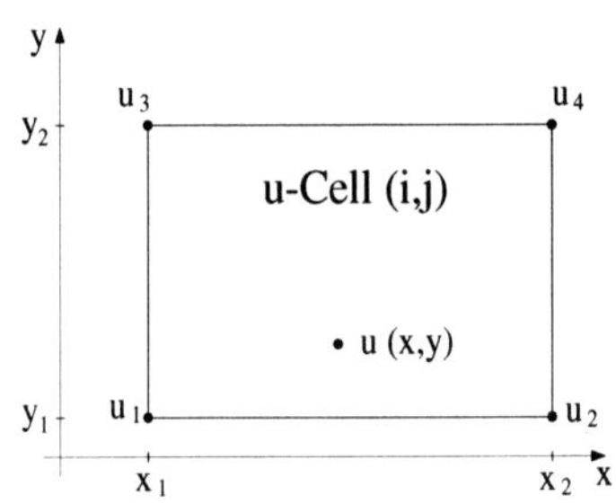

FIG. 4.4. *u-cell (i,j).*

are available. We can now approximate the desired velocity $u(x,y)$ at position (x,y) using bilinear interpolation:

$$u(x,y) = \frac{1}{\delta x\,\delta y}\Big[(x_2 - x)(y_2 - y)u_1 + (x - x_1)(y_2 - y)u_2 + (x_2 - x)(y - y_1)u_3 + (x - x_1)(y - y_1)u_4 \Big].$$

2. *Computation of $v(x,y)$:* The procedure here is analogous to 1: The indices of the v-cell surrounding the position (x,y) are

$$i := (int)\left(\frac{x + \frac{\delta x}{2}}{\delta x}\right) + 1, \quad j := (int)\left(\frac{y}{\delta y}\right) + 1;$$

the positions of the corners containing the v-velocities in the staggered grid are

$$x_1 := ((i-1) - \tfrac{1}{2})\,\delta x, \qquad y_1 := (j-1)\,\delta y,$$

$$x_2 := (i - \tfrac{1}{2})\,\delta x, \qquad y_2 := j\,\delta y,$$

and the interpolation scheme reads

$$v(x,y) = \frac{1}{\delta x\,\delta y}\Big[(x_2 - x)(y_2 - y)v_1 + (x - x_1)(y_2 - y)v_2 + (x_2 - x)(y - y_1)v_3 + (x - x_1)(y - y_1)v_4\Big]$$

with the four staggered grid velocities

$$v_1 := v_{i-1,j-1}\,, \quad v_2 := v_{i,j-1}\,, \quad v_3 := v_{i-1,j}\,, \quad v_4 := v_{i,j}\,.$$

4.2.2 Implementation

Two new functions, PARTICLE_TRACING and STREAKLINES, must be written. These are to be called from within the time-stepping loop of the main program to compute the pathlines and streaklines for given positions as described above and to write these out to a file. In the case of streaklines, the data on these files can then be displayed as a sequence of images to show the time evolution using suitable visualization routines. The particles are to be kept in the following data structure:

```
struct particle{
        REAL x, y;
        struct particle *next;
}

struct particleline{
        int length;
        struct particle *Particles;
}
struct particleline *Particlelines;
```

A schematic representation of the data structure `Particlelines` is shown in Figure 4.5, from which we may also infer the definition of the `length` of a `particleline`.

If we use this data structure to store a streakline, then the zeroth element of a `particleline` belonging to `Particlelines` serves as an auxiliary element for storing each particle position where particles are to be injected at given times. The particles which have been moved along in previous time steps are stored in the list elements that follow. The `particleline` lists within `Particlelines` thus grow in length over the course of the simulation. When we use the data structure `Particlelines` to trace particles, all positions of the particles being tracked can be held in one `particleline`. After each time step, the position of each particle is overwritten with the updated location, so that, in this case, the length of a `particleline` list remains constant.

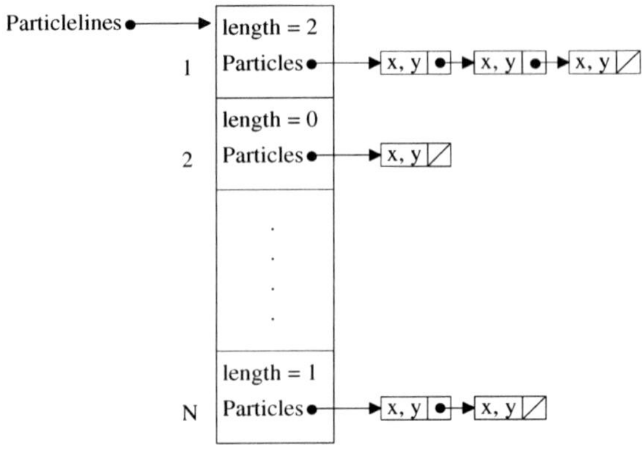

FIG. 4.5. *Data structure* `Particlelines` *with examples for* particleline *lengths.*

The required new functions are as follows.

1. `struct particleline *SET_PARTICLES(int N,REAL x1,REAL y1,REAL x2, REAL y2)`: Memory for the particles is allocated dynamically and the particle positions are entered in each `particleline`. Each particle line initially has length 1 (see Figure 4.5). The auxiliary element and the first particle contain the same coordinates. A pointer to the allocated region is returned.
   The particles are to be injected along the line segment with end points (`x1`,`y1`) and (x2,y2); the parameter `N` determines the number of particles to be uniformly distributed along this line.

   Note. If one sets `Particlelines=SET_PARTICLES(N,x1,y1,x2,y2)` for

   - `PARTICLE_TRACING`, then the first element of each `particleline` is merely a dummy element,
   - `STREAKLINES`, then the first element of each `particleline` contains the position where the particle is to be injected, while all following elements contain the positions of the actual particles to be traced. In contrast with the function `PARTICLE_TRACING`, additional particles are inserted into the list in the course of the computation.

   The parameters `N,x1,y1,x2,y2` should be read from the input file in the function `READ_PARAMETER`. The call to this function occurs in `main`.

2. `void ADVANCE_PARTICLES(imax,jmax,delx,dely,delt,U,V,int N,struct particleline *Partlines)`: The particle positions of time t_n held in `Partlines` (an array of `particleline`) are advanced to the particle positions of time t_{n+1} according to (4.3). Those particles which, as a result, move out of the base domain $\mathcal{G} = [0, i_{\max}\,\delta x] \times [0, j_{\max}\,\delta y]$ or into obstacle cells are deleted from the list, and the memory occupied by these particles is returned using `free`.[37]

[37] As long as the CFL condition is satisfied, particles at free-slip or no-slip boundaries can at most unlawfully disappear into corner cells, but not into edge cells. If this is to be prevented, then the interpolation algorithm needs to be modified in the vicinity of edge cells by using the additional information that the normal velocity component vanishes not only at the midpoints of the boundary edges but also along the entire edge.

The call to this function occurs from PARTICLE_TRACING and STREAKLINES.

3. INJECT_PARTICLES(int N,struct particleline *Partlines): The particles initialized in SET_PARTICLES are injected along the given line segment by inserting each particle at the front of each particleline in Partlines with the position specified in the dummy element. In addition, storage must be allocated for the new particle. This function is called in STREAKLINES.

4. void WRITE_PARTICLES(char* outputfile,int N,struct particleline *Partlines): The particle positions contained in Partlines are written to the file specified in the input parameter outputfile.
In order that this function may be used by both functions PARTICLE_TRACING and STREAKLINES, it is recommended that the file outputfile be opened for writing with the statement fopen(outputfile,"a"). If this file doesn't exist, this statement creates it; otherwise, it appends the particle positions to be written at the end of the file.

5. void PARTICLE_TRACING(char* outputfile,REAL delt_trace,t,imax,jmax, delx, dely, delt, U, V, int N, struct particleline *Partlines): Uses ADVANCE_PARTICLES to compute the current positions of the particles to be traced at each time step and overwrites their old positions in Partlines. At time intervals delt_trace, the current particle positions are appended to the file specified in the input parameter outputfile using WRITE_PARTICLES. This file contains the quantities imax, jmax, delx, dely, N as a header.[38]
Note. It must be ensured that the file outputfile contains no data from previous runs. The output format must also be adapted to the visualization tool being used.
This function is called in the time-stepping loop in main.

6. void STREAKLINES(char* outputfile,REAL delt_inject,REAL delt_streak, t, imax, jmax, delx, dely, delt, U, V, int N, struct particleline *Partlines): At each time step, the current particle positions are calculated with the help of ADVANCE_PARTICLES and their old positions in Partlines are overwritten. At time intervals of delt_inject, particles are injected and, at intervals of delt_streak, the current particle positions are written to the file specified in the input parameter outputfile by way of the function WRITE_PARTICLES.[38]
The specific output of the particle positions in WRITE_PARTICLES must be adapted to the visualization tool being used. Data written at different time instances should also be visualized at separate times.
The call to this function occurs in the time-stepping loop of main.

[38]Since the length of each time step is determined by a stepsize control scheme, the values of delt_trace, delt_inject, and delt_streak cannot be a priori determined such that their scheduling is synchronized with the time steps. Usually, however, delt is small compared to delt_trace, delt_inject, and delt_streak, so each action can be performed the first time its scheduled time is exceeded.

These functions can be added to the module `visual.c`.

4.3 Stream Function and Vorticity

Having introduced special techniques for visualizing the state and the evolution of flows in the two previous sections, we now turn our attention to two new physical quantities which describe certain characteristics of the flow. We first define these and give their physical interpretation. Moreover, we show that the Navier–Stokes equations in two space dimensions ((2.1) or (2.2a,b)) in the *primitive variables* u, v, p may also be completely described in terms of these new quantities, the *stream function* ψ and the *vorticity* ζ.

4.3.1 Definition and Interpretation

If u and v are the velocities of the flow field, then the *stream function* $\psi(x,y)$ is defined by

$$\frac{\partial\psi(x,y)}{\partial x} := -v, \qquad \frac{\partial\psi(x,y)}{\partial y} := u, \tag{4.4}$$

and the *vorticity* by

$$\zeta(x,y) := \frac{\partial u}{\partial y} - \frac{\partial v}{\partial x}. \tag{4.5}$$

These quantities may be interpreted as follows.

1. *The vorticity* ζ: ζ measures—as the name suggests—the strength of vortical motion in the velocity field. For an idealization, we turn to Figure 4.6.

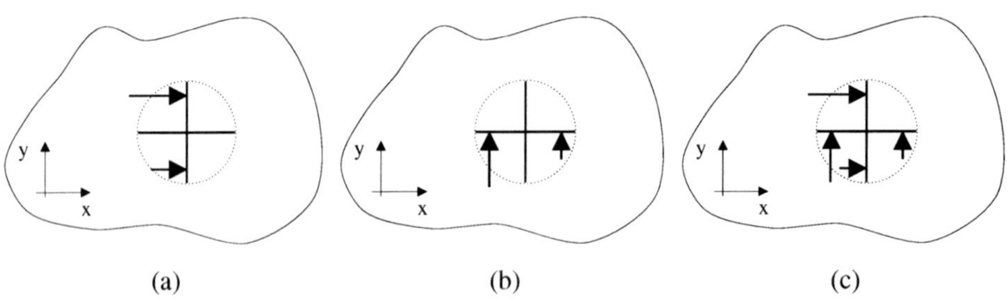

FIG. 4.6. *Thought experiment illustrating the vorticity* ζ.

An imaginary wheel has been inserted into the velocity field. In case (a), the u-velocity *increases* in the y-direction, $\partial u/\partial y > 0$, causing the wheel to rotate clockwise. In case (b), clockwise rotation is also produced, but this time due to the v-velocity *decreasing* in the x-direction: $\partial v/\partial x < 0$.
To account for the total rotational effect, cases (a) and (b) are combined in case (c), and we obtain the vorticity $\zeta = \partial u/\partial y - \partial v/\partial x$.

2. *Stream function ψ:* For a function of two variables x and y to be well defined by its partial derivatives as in (4.4), a sufficient condition known as the *integrability condition* (interchangeability of the order of derivatives) must be satisfied:

$$\frac{\partial^2\psi}{\partial x\partial y} = \frac{\partial^2\psi}{\partial y\partial x}\,. \tag{4.6}$$

In the case of the stream function, it is the continuity equation (2.2c) which ensures that this condition holds, i.e.,

$$\frac{\partial}{\partial x}u + \frac{\partial}{\partial y}v = 0 \iff \frac{\partial}{\partial x}\left(\frac{\partial\psi}{\partial y}\right) + \frac{\partial}{\partial y}\left(-\frac{\partial\psi}{\partial x}\right) = 0$$

$$\iff \frac{\partial^2\psi}{\partial x\partial y} - \frac{\partial^2\psi}{\partial y\partial x} = 0\,.$$

For a physical interpretation of the stream function, we first introduce the concept of a *streamline.*

Definition. A *streamline* is a curve whose tangent is parallel to the velocity vector $\binom{u}{v}$ in each of its points (x, y) at a fixed time t.

The relationship between the stream function ψ and streamlines of the velocity field $\binom{u}{v}$ lies in the fact that the streamlines are the contour lines of ψ. This is seen as follows.

i. Along a contour line $\varphi_c(s)$ of ψ, the gradient of ψ is orthogonal to the tangential directions $\dot\varphi_c(s)$, as follows from (4.2):

$$0 = \frac{d\psi(\varphi_c(s))}{ds} = \frac{\partial\psi}{\partial x}\frac{dx_c(s)}{ds} + \frac{\partial\psi}{\partial y}\frac{dy_c(s)}{ds}$$
$$= \left(\frac{\partial\psi}{\partial x}, \frac{\partial\psi}{\partial y}\right)\cdot\begin{pmatrix}\dot x_c(s)\\ \dot y_c(s)\end{pmatrix} = \mathrm{grad}\psi\cdot\dot\varphi_c(s)\,.$$

ii. The gradient of the stream function ψ is orthogonal to the velocity field:

$$\mathrm{grad}\psi\cdot\begin{pmatrix}u\\ v\end{pmatrix} = \left(\frac{\partial\psi}{\partial x}, \frac{\partial\psi}{\partial y}\right)\cdot\begin{pmatrix}u\\ v\end{pmatrix} = (-v, u)\cdot\begin{pmatrix}u\\ v\end{pmatrix} = -uv + uv = 0\,.$$

iii. From i and ii we now conclude that the tangential directions $\dot\varphi_c(s)$ and the velocity vectors are parallel along the level curves. Thus the contour lines of the stream function satisfy the definition of a streamline.

Another property of the stream function is that the mass flow rate m^t between two streamlines $\psi(x, y) = \psi_a$, $\psi(x, y) = \psi_b$ is a constant given by

$$m^t = \psi_b - \psi_a.$$

The mass flow rate m^t is defined as the mass passing through a curve c connecting two points a and b on the streamlines ψ_a and ψ_b per unit time (see Figure 4.7). In general, the mass m is given by the integral $m = \int_\Omega \varrho\, dxdy$, where ϱ is the density of the fluid, and the mass passing through a curve c is given by

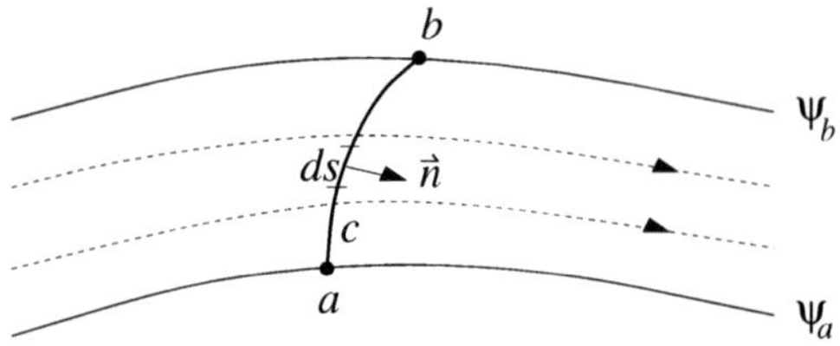

FIG. 4.7. *Mass flow between streamlines; c denotes a curve connecting points a and b on the streamlines ψ_a and ψ_b, $\vec{n}$ is the unit normal vector, and s is the arc length of the curve c.*

$$m^t = \int_c \varrho \begin{pmatrix} u \\ v \end{pmatrix} \cdot \vec{n}\, ds\,.$$

Here $\vec{n}$ is the normal to c in the flow direction and s denotes a parametrization of the curve c (see Figure 4.7).

For incompressible flows, we have $\varrho = \varrho_0 = \text{const}$ and, normalizing $\varrho_0 = 1$, we obtain

$$\begin{aligned} m^t &= \int_c \varrho_0 \begin{pmatrix} u \\ v \end{pmatrix} \cdot \vec{n}\, ds = \int_c \begin{pmatrix} u \\ v \end{pmatrix} \cdot \vec{n}\, ds = \int_a^b \begin{pmatrix} u \\ v \end{pmatrix} \cdot \begin{pmatrix} dy \\ -dx \end{pmatrix} = \int_a^b u dy - v dx \\ &= \int_a^b \frac{\partial \psi}{\partial y} dy + \frac{\partial \psi}{\partial x} dx = \int_a^b d\psi = \psi_b - \psi_a\,. \end{aligned}$$

Thus, in incompressible flows, the mass flowing between two streamlines is always the same. We also note that, for stationary flows ($\partial/\partial t = 0$), the pathlines and streaklines introduced in Section 4.2 coincide with the streamlines defined here.

4.3.2 Implementation

The quantities to be computed are the discrete values $\psi_{i,j}$ and $\zeta_{i,j}$. To avoid tedious interpolations in the staggered grid, these quantities are to be computed in the *upper right corner* of each cell (i,j) rather than in the cell centers. We assign the indices $i = 0, \ldots, i_{\max}$, $j = 0, \ldots, j_{\max}$ to the stream function $\psi_{i,j}$, and $i = 1, \ldots, i_{\max} - 1$, $j = 1, \ldots, j_{\max} - 1$ to the vorticity values $\zeta_{i,j}$. The new data arrays required are

`REAL`	`PSI`	stream function ψ,
`REAL`	`ZETA`	vorticity ζ,

and we need to implement the following functions.

1. `void COMP_PSI_ZETA(U,V,REAL **PSI,REAL **ZETA,int **FLAG,imax, jmax,delx,dely)`: Computes the discrete stream function $\psi_{i,j}$ using the defining equation

$$\frac{\partial \psi}{\partial y} = u \quad \text{via the discretization} \quad \left[\frac{\partial \psi}{\partial y}\right]_{i,j} = \frac{\psi_{i,j} - \psi_{i,j-1}}{\delta y}$$

by fixing $\psi_{i,0} := 0$ and using the formula

$$\psi_{i,j} = \psi_{i,j-1} + u_{i,j}\,\delta y\,, \quad i = 0,\ldots,i_{\max}\,, \quad j = 1,\ldots,j_{\max}\,;$$

the discrete vorticity $\zeta_{i,j}$ is calculated using the defining equation (4.5) according to the formula

$$\zeta_{i,j} := \frac{u_{i,j+1} - u_{i,j}}{\delta y} - \frac{v_{i+1,j} - v_{i,j}}{\delta x}\,.$$

In each case, the obstacle structure specified by the flag array must be accounted for: $\zeta_{i,j}$ can be set to zero inside obstacle cells and on their boundaries, and for $\psi_{i,j}$, no additions are performed inside obstacle cells; i.e., $\psi_{i,j} = \psi_{i,j-1}$ there.

2. The function `OUTPUTVEC` must be augmented by the output of the arrays `PSI` and `ZETA`.

The new function `COMP_PSI_ZETA` can be incorporated into the module `visual.c`.

4.3.3 The Stream Function–Vorticity Formulation of the Navier–Stokes Equations

In two dimensions, the Navier–Stokes equations are often expressed using the stream function ψ and the vorticity ζ in place of the primitive variables u, v, and p. This involves the elimination of the pressure p, thus yielding one dependent variable less. In three dimensions, however, this formulation leads to six unknowns rather than four (in primitive variables), which makes this approach less attractive for that case.[39]

In the following, we briefly derive the resulting two-dimensional equations for ψ and ζ. We begin by considering the momentum equations in (2.1):

$$\begin{aligned}
\frac{\partial u}{\partial t} + u\frac{\partial u}{\partial x} + v\frac{\partial u}{\partial y} + \frac{\partial p}{\partial x} &= \frac{1}{Re}\left(\frac{\partial^2 u}{\partial x^2} + \frac{\partial^2 u}{\partial y^2}\right) + g_x \qquad \Big|\ \frac{\partial}{\partial y},\\
\frac{\partial v}{\partial t} + u\frac{\partial v}{\partial x} + v\frac{\partial v}{\partial y} + \frac{\partial p}{\partial y} &= \frac{1}{Re}\left(\frac{\partial^2 v}{\partial x^2} + \frac{\partial^2 v}{\partial y^2}\right) + g_y \qquad \Big|\ \frac{\partial}{\partial x},
\end{aligned}$$

from which we eliminate the pressure by differentiating the first equation by y and the second by x, subtracting the second from the first, and then substituting definition (4.5) for the vorticity $\zeta = \partial u/\partial y - \partial v/\partial x$. In this manner, we obtain the equation

$$\frac{\partial \zeta}{\partial t} + \frac{\partial u}{\partial x}\zeta + u\frac{\partial \zeta}{\partial x} + \frac{\partial v}{\partial y}\zeta + v\frac{\partial \zeta}{\partial y} = \frac{1}{Re}\left(\frac{\partial^2 \zeta}{\partial x^2} + \frac{\partial^2 \zeta}{\partial y^2}\right) + \left(\frac{\partial g_x}{\partial y} - \frac{\partial g_y}{\partial x}\right),$$

[39] A detailed derivation of the two- and three-dimensional equations is found, e.g., in [Fletcher, 1991].

which we first rid of the terms containing $\partial u/\partial x$ and $\partial v/\partial y$ using the continuity equation and finally, by using the defining equation (4.4) for the stream function $\partial\psi/\partial y = u$, $\partial\psi/\partial x = -v$, transform into what is known as the *vorticity transport equation*:

$$\frac{\partial}{\partial t}\zeta + \frac{\partial\psi}{\partial y}\frac{\partial\zeta}{\partial x} - \frac{\partial\psi}{\partial x}\frac{\partial\zeta}{\partial y} = \frac{1}{Re}\Delta\zeta + \left(\frac{\partial g_x}{\partial y} - \frac{\partial g_y}{\partial x}\right). \tag{4.7}$$

Another equation is obtained by inserting the definition (4.4) of the stream function into that of the vorticity (4.5). This equation is known as the *Poisson equation for the stream function*

$$\Delta\psi = \zeta. \tag{4.8}$$

The two equations (4.7) and (4.8) form a nonlinear coupled system of equations in which pressure has been eliminated and in which the continuity equation is satisfied automatically.

This formulation is often more efficient than that using the primitive variables, since introducing the stream function eliminates the explicit solution of the continuity equation and since it contains only two unknowns (ψ and ζ) instead of three (u, v, and p). The treatment of the boundary conditions, however, becomes more complicated. The first and often used discretization at the boundary for this case goes back to [Thom, 1933]. The stream function–vorticity formulation is also used for numerical computations in [Mallinson & de Vahl Davis, 1973], [Roache, 1976], and [Ghia et al., 1982].

The extension to the three-dimensional case is also not as simple and straightforward as for the primitive variable formulation, as it requires a suitable generalization of the definitions of vorticity ζ and stream function ψ. Moreover, the advantage regarding storage space is lost since, as mentioned, six functions must be solved for instead of two. Numerical computations for three-dimensional problems using this formulation are performed in the works of [Aziz & Hellums, 1967], [Hirasaki & Hellums, 1970], [Mallinson & de Vahl Davis, 1977], and [Wong & Reizes, 1984].

Besides the stream function–vorticity formulation of the Navier–Stokes equations given above, there is also the vorticity-velocity formulation, in which pressure is eliminated as well. In two dimensions, it reads

$$\begin{aligned}
\frac{\partial}{\partial t}\zeta + u\frac{\partial\zeta}{\partial x} + v\frac{\partial\zeta}{\partial y} &= \frac{1}{Re}\Delta\zeta + \left(\frac{\partial g_x}{\partial y} - \frac{\partial g_y}{\partial x}\right), \\
\frac{\partial u}{\partial x} + \frac{\partial v}{\partial y} &= 0, \\
\zeta(x,y) &= \frac{\partial u}{\partial y} - \frac{\partial v}{\partial x}.
\end{aligned} \tag{4.9}$$

A derivation for the two- and three-dimensional cases is also found in [Fletcher, 1991]. Since the stream function ψ is not used here, the continuity equation must once more be treated separately. For the extension to three dimensions, the definition of the vorticity ζ must again be appropriately generalized, and the number of unknowns again increases to six (three velocities and three vorticity components) as opposed to four in the primitive variables formulation (u, v, w, and p). The treatment of the boundary is once again made more difficult by the vorticity ζ. Numerical experiments based on this formulation can be found, e.g., in [Richardson & Cornish, 1977], [Dennis et al., 1979], [Dennis, 1985], [Gatski et al., 1982], [Gatski & Grosch, 1985], and [Gatski et al., 1989].

Chapter 5

Example Applications

In this chapter we collect some results calculated using the program developed in the last two chapters. We wish to emphasize the diversity of the possible applications and to encourage readers to perform further experiments on their own.

We begin with three classical CFD problems: the lid-driven cavity, the flow over a backward-facing step, and the flow past an obstacle. We include the flow in a T-shaped pipe junction as an example with time-dependent boundary conditions and the flow through porous media as an example involving a highly complex geometry. Finally, we consider a micropump as an example of fluid-structure interaction.

Photographs of the associated physical experiments can be found, e.g., in the picture albums [Nakayama, 1988] and [van Dyke, 1982]. We refer also to the films [Prandtl, 1936a] and [Prandtl, 1936b].

5.1 Lid-Driven Cavity

As a first example, for which we can still do without the flag array, we simulate a *driven cavity flow* in a square domain (cf. also [Ghia et al., 1982], [Pinelli & Vacca, 1994]). The physical configuration (Figure 5.1) consists of a square container filled with a fluid. The lid of the container moves at a given, constant velocity, thereby setting the fluid in motion.

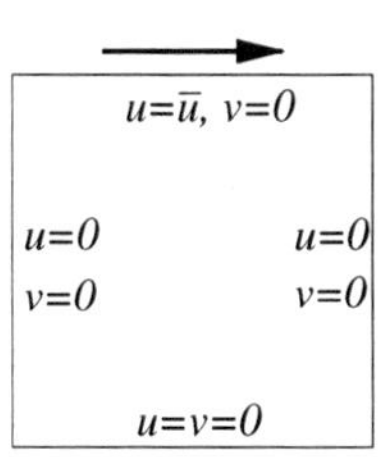

FIG. 5.1. *Driven cavity flow, problem configuration.*

No-slip conditions are imposed on all four segments of the boundary with the exception of the upper boundary, along which the velocity u in the x-direction is not set to zero, but equal to the given lid velocity $\bar{u}$ to simulate the moving lid. In our program this is implemented by setting the boundary values along the upper boundary to

$$u_{i,j_{\max}+1} = 2.0\,\bar{u} - u_{i,j_{\max}}, \quad i = 1, \ldots, i_{\max}.$$

The specific parameters used in our calculations are as follows:

```
imax = 128,     jmax = 128,   xlength = 1.0,   ylength = 1.0,
delt = 0.02,    tau = 0.5,
eps = 0.001,    omg = 1.7,    gamma = 0.9,     itermax = 100,
GX = 0.0,       GY = 0.0,     Re = 1000,
UI = 0.0,       VI = 0.0,     PI = 0.0,
wW = 2,         wE = 2,       wN = 2,          wS = 2.
```

The following pages (Figures 5.2 to 5.5) show the velocity field and the streamlines for $Re = 1000$, both evolving in time as well as at steady state. At $t = 0$, the lid velocity $\bar{u}$ is instantaneously set from zero to one, thereby slowly setting in motion the fluid initially at rest. The formation of the large primary eddy as well as that of the first counter-rotating secondary eddy in the lower right corner can be observed rather well. The counter-rotating eddy in the lower left corner, however, takes considerably longer to develop.

Next we show the steady state velocity field, streamlines, and contour lines of vorticity at different Reynolds numbers in Figures 5.6 through 5.8.[40] Figure 5.7 clearly shows that the size of the first counter-rotating eddies depends on the Reynolds number. At Reynolds numbers above 1000, we can also observe the second counter-rotating eddies in the lower corners. It can be shown analytically that for $Re = 0$ an infinite sequence of eddies and countereddies is established whose size decreases exponentially [Moffatt, 1964]. At our level of discretization, however, we are only able to resolve the first two eddies. A numerical search for further eddies would require adaptive refinement of the grid in the lower corners, as is demonstrated in [Reichert & Wittum, 1994], in which, up to the level of computing accuracy, the smaller and smaller eddies in the corners are revealed by magnifying the images.

Since the flow in this problem ultimately reaches steady state, the explicit time-stepping scheme we have described is clearly inferior to implicit methods, as the latter allow much larger time steps to be taken. In [Ghia et al., 1982] the steady Navier–Stokes equations in the stream function–vorticity formulation are actually used. This approach, however, does not reveal the time evolution of the flow after the lid is set in motion.

[40]The contour levels of ψ and ζ are those used in [Ghia et al., 1982].

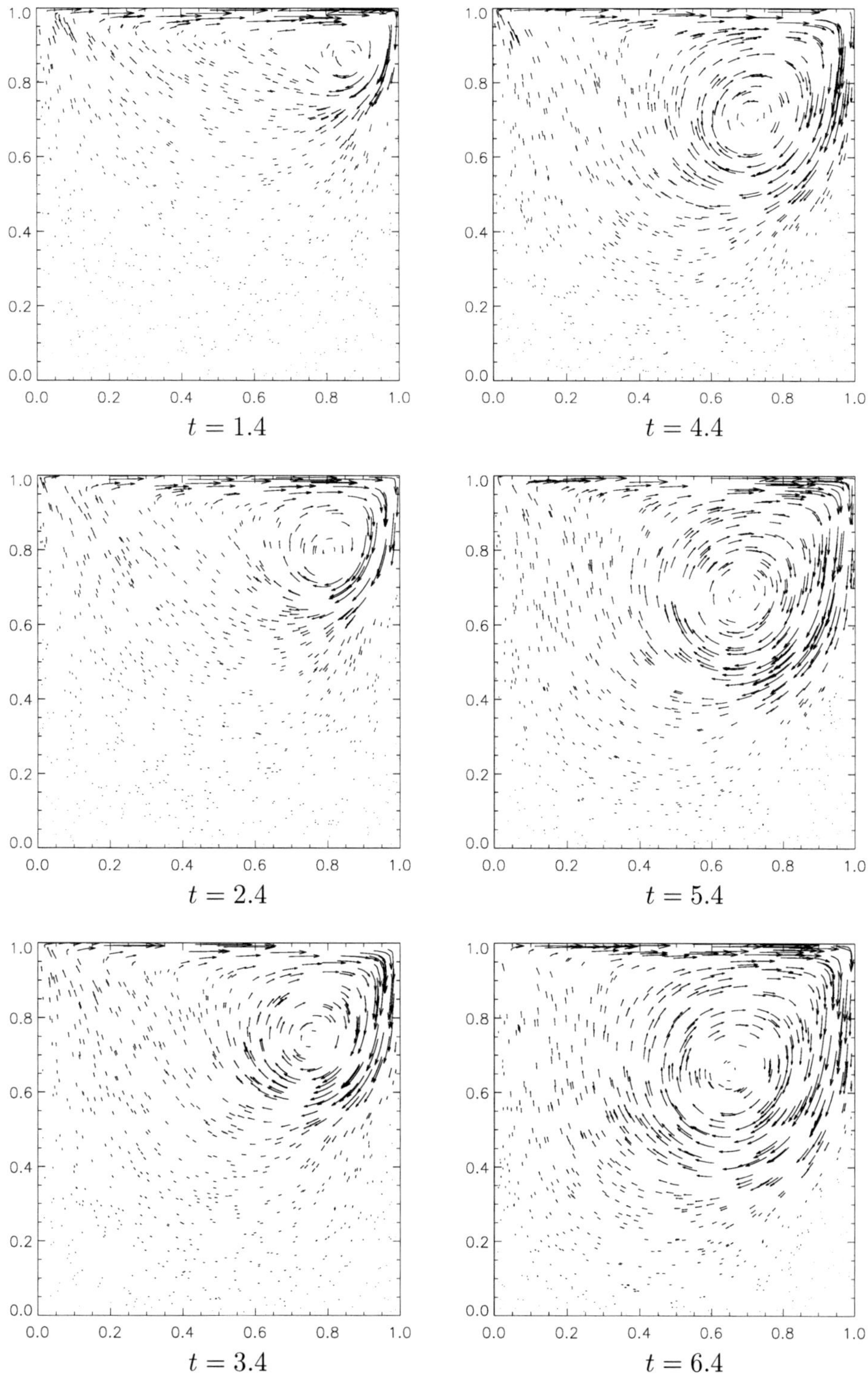

FIG. 5.2. *Driven cavity, velocity field, time evolution at* $Re = 1000$.

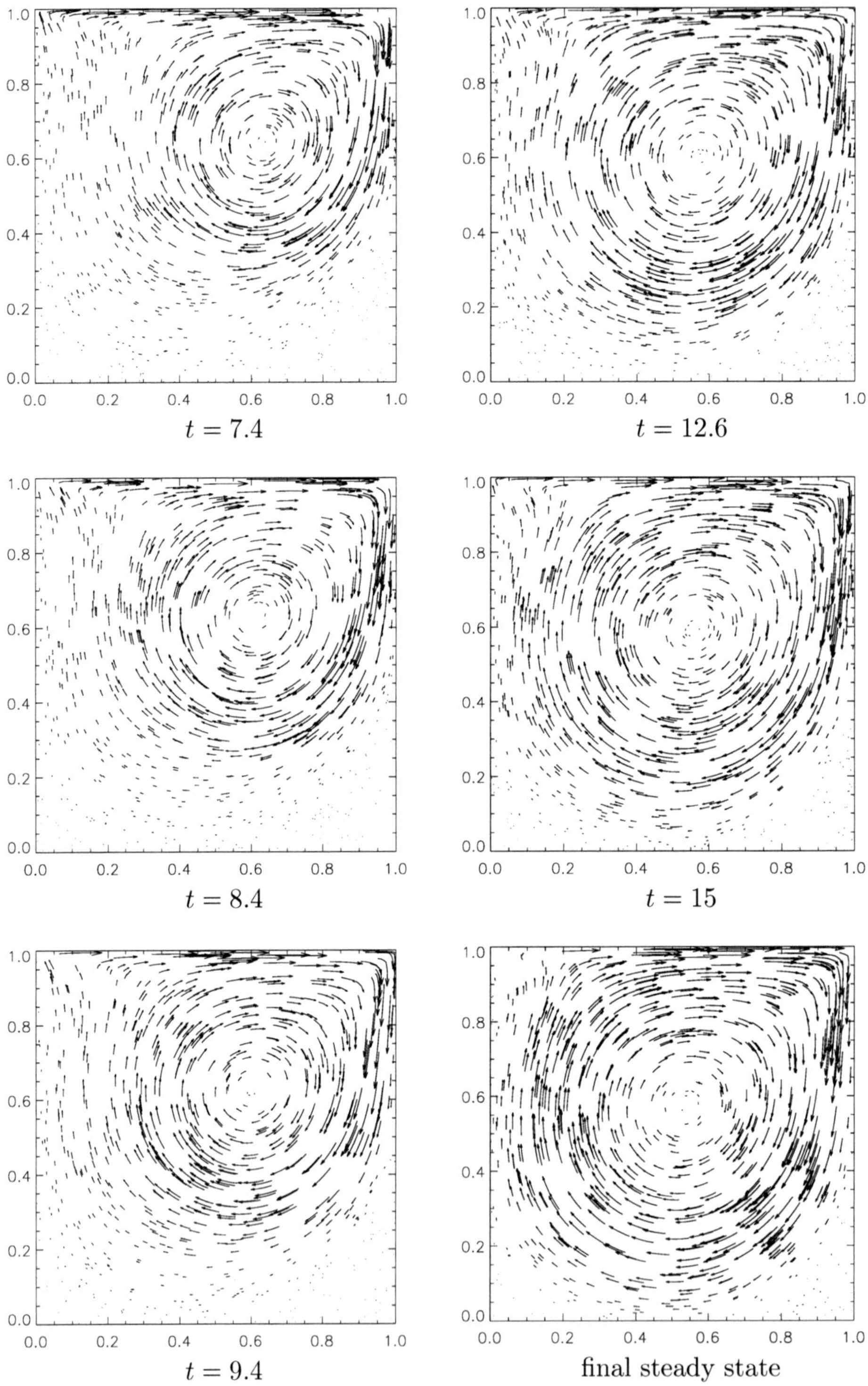

FIG. 5.3. *Driven cavity, velocity field, time evolution at* $Re = 1000$.

FIG. 5.4. *Cavity flow, streamlines, time evolution at* $Re = 1000$.

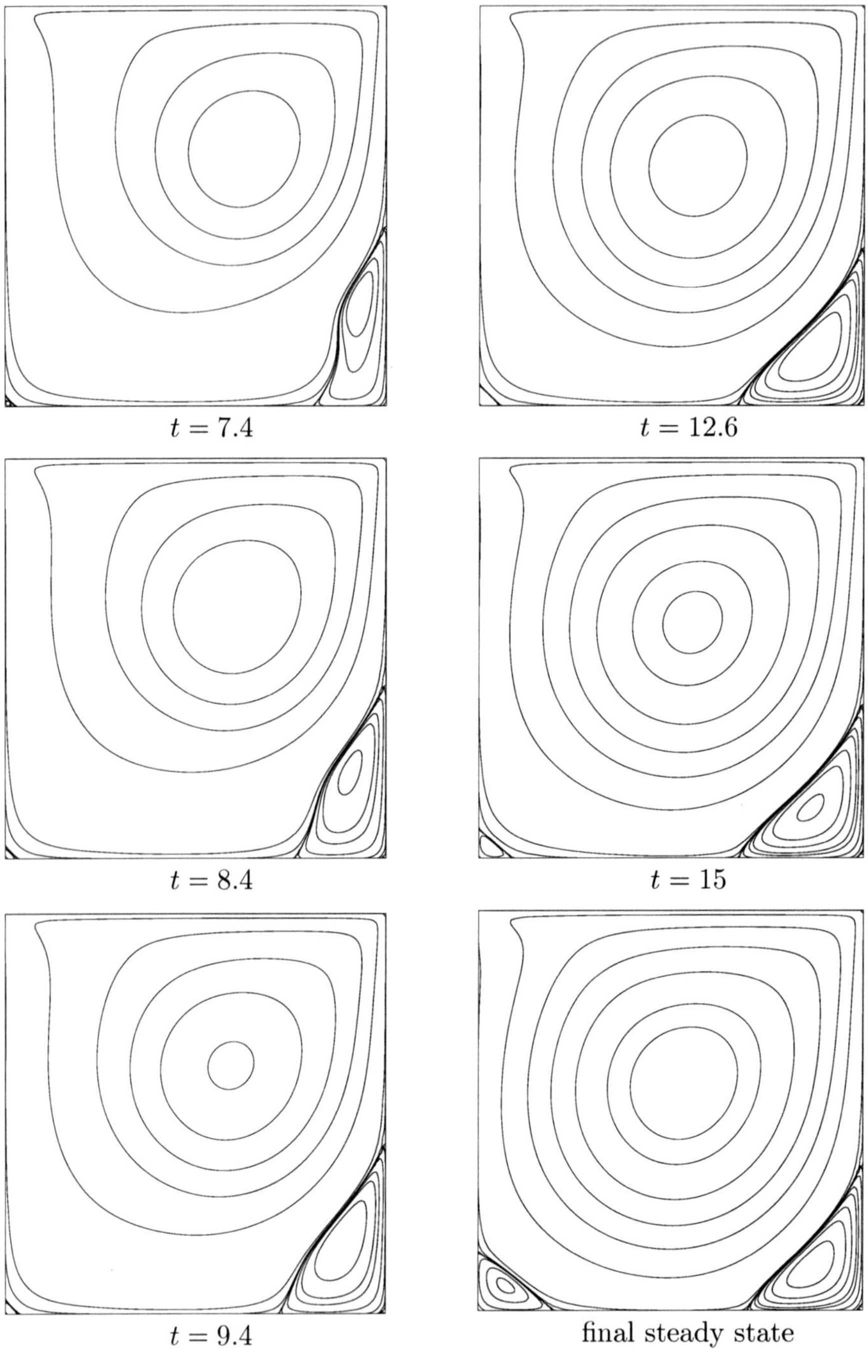

FIG. 5.5. *Driven cavity, streamlines, time evolution at* $Re = 1000$.

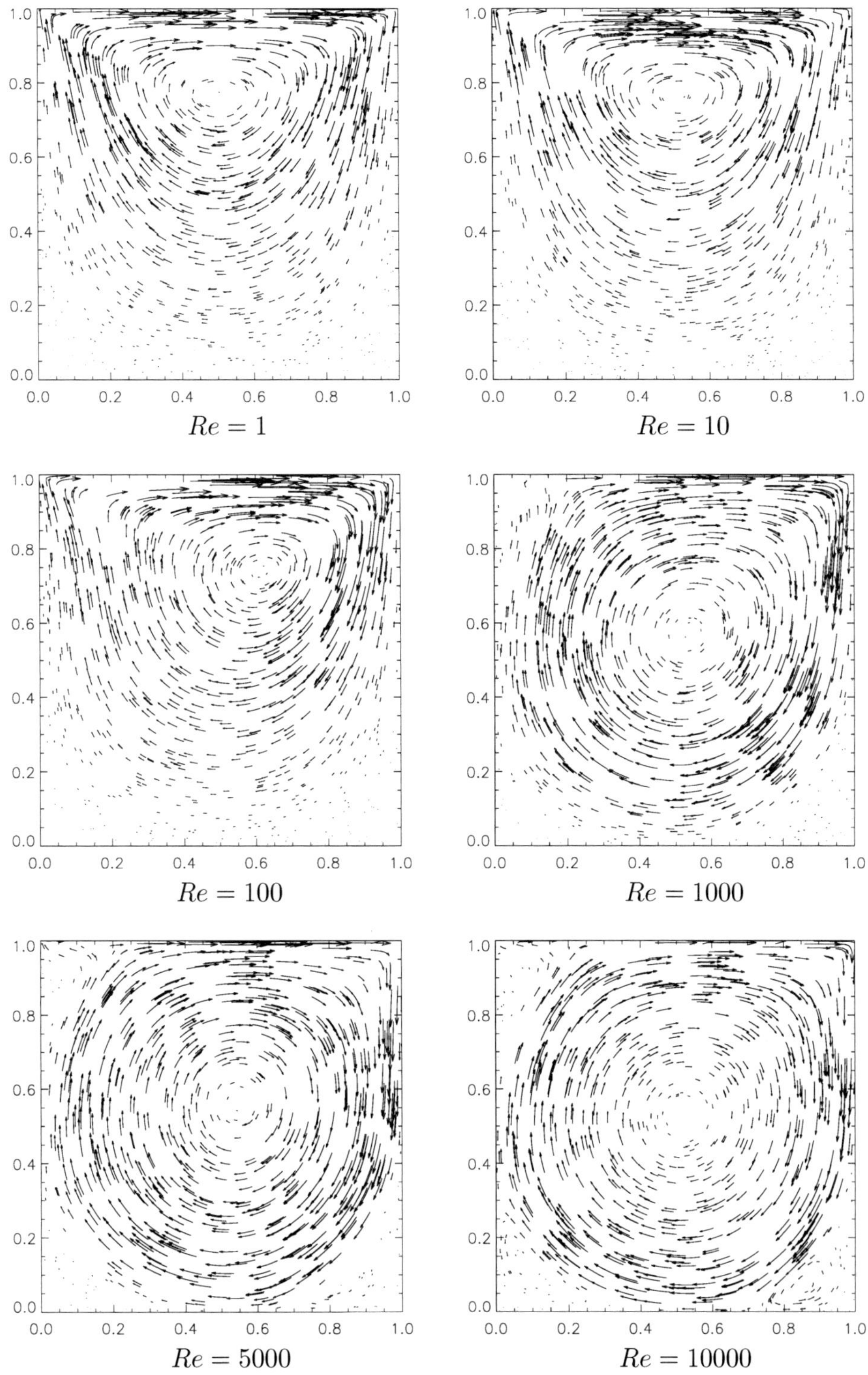

FIG. 5.6. *Driven cavity, steady state velocity field at different Reynolds numbers.*

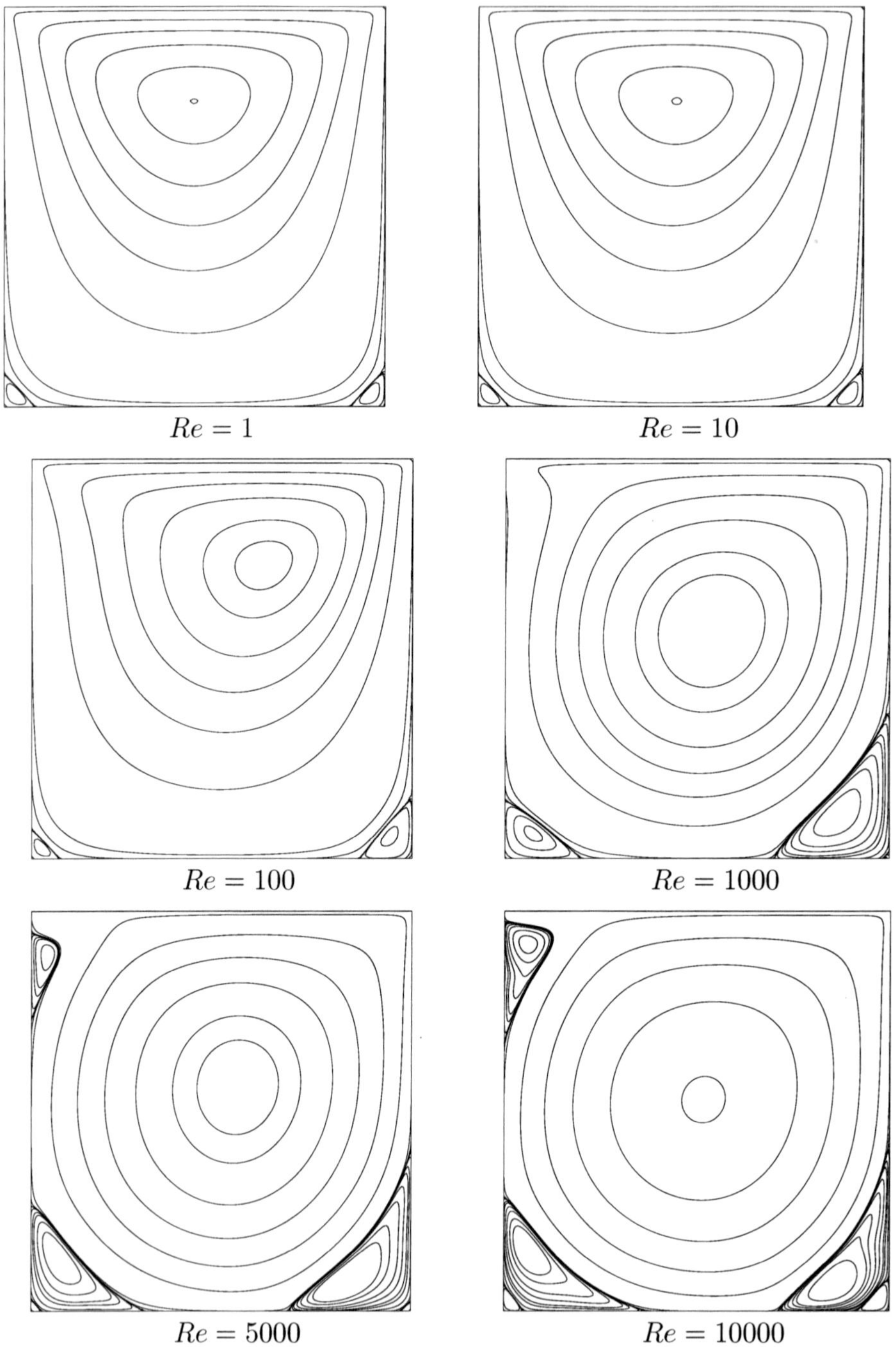

FIG. 5.7. *Driven cavity, steady state streamlines at different Reynolds numbers.*

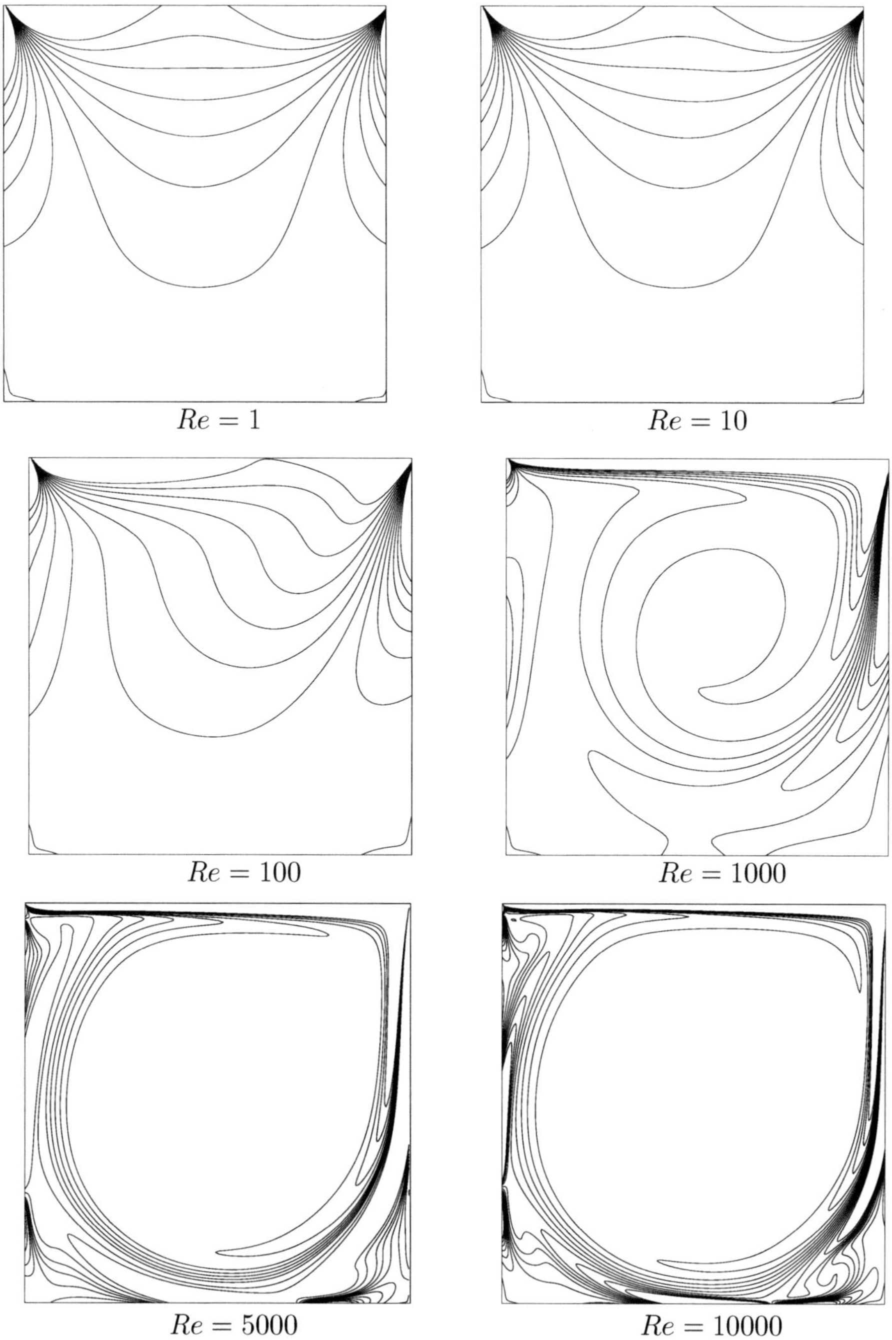

FIG. 5.8. *Driven cavity, steady state vorticity at different Reynolds numbers.*

There are also many problems in which the boundary conditions change with time (in this case, the lid velocity $\bar{u}$), so that no steady state solution exists. An example is a lid oscillating in time at a frequency f and with a velocity amplitude $\bar{u}_0$ for which $\bar{u}(t) = \bar{u}_0 \sin(2\pi f\, t)$, as is treated in [Durst, 1991]. An application could, for instance, be an oscillating quartz or a piezo-crystal exciting a fluid through its oscillation. At higher oscillation frequencies an implicit method may lose its advantages since the oscillations of the lid must be resolved by increasingly small time steps. We leave numerical experiments to the reader here.

5.2 Flow over a Backward-Facing Step

As a second example, we consider the flow over a *backward-facing step*, which has become popular as a test problem for developing flow simulation codes. It consists of a fluid flowing in a straight channel which abruptly widens on one side. Numerical results obtained using a wide range of methods can be found in [Gartling, 1990], [Kaiktsis et al., 1991], [Kim & Moin, 1985], [Sohn, 1988], and [Turek, 1992] among others. Results of physical experiments are given in [Armaly et al., 1983], for example (cf. also Figure 113 in [Nakayama, 1988]).

The parameters we have used are

```
imax = 300,    jmax = 75,   xlength = 30,   ylength = 1.5,
delt = 0.02,   tau = 0.5,
eps = 0.001,   omg = 1.7,   gamma = 0.9,    itermax = 100,
GX = 0.0,      GY = 0.0,
UI = 1.0,      VI = 0.0,    PI = 0.0,
wW = 3,        wE = 3,      wN = 2,         wS = 2.
```

The obstacle domain representing the step is the rectangle $[0, 7.5] \times [0, 0.75]$ and the inflow velocity at the left boundary has the constant value $u_{\text{in}} = 1.0$. In the function `INIT_FLAG` the flags of all cells in this obstacle region are thus to be set to the value `C_B`, whereas all remaining cells receive the flag value `C_F`. When setting the initial values of u it is useful to only set $u_0 = 1.0$ in the upper half of the domain, leaving $u_0 = 0.0$ in the lower half. This ensures that the initial velocity satisfies the discrete continuity equation. Note also that the following figures show only the part of the computational domain between $x = 5.0$ and $x = 20.0$, since this contains all the essential features. The actual computational domain was chosen larger in order to obtain a parabolic velocity profile at the step and at the outflow boundary.

The streamlines shown in Figure 5.9 reveal that, for small Reynolds numbers ($Re = 1$), the flow widens immediately behind the step and only a very small eddy is formed. This eddy increases in size with increasing Reynolds number. When viscosity is further reduced ($Re = 250, 500$), the main flow is drawn downward, ultimately causing it to separate from the upper boundary and leading to the formation of a second eddy there. The lengths x_1 and x_2 of the upper and lower eddies as well as the horizontal distance x_3 from the step to the upper eddy's point of separation—each normalized by the step height s—are values often used in CFD to characterize the resulting flow (cf. [Armaly et al., 1983]). In our simulations we have obtained the following values.

Re	x_1/s	x_2/s	x_3/s
1	0.3	—	—
100	3.8	—	—
250	5.8	5.5	4.4
500	8.3	9.1	6.2

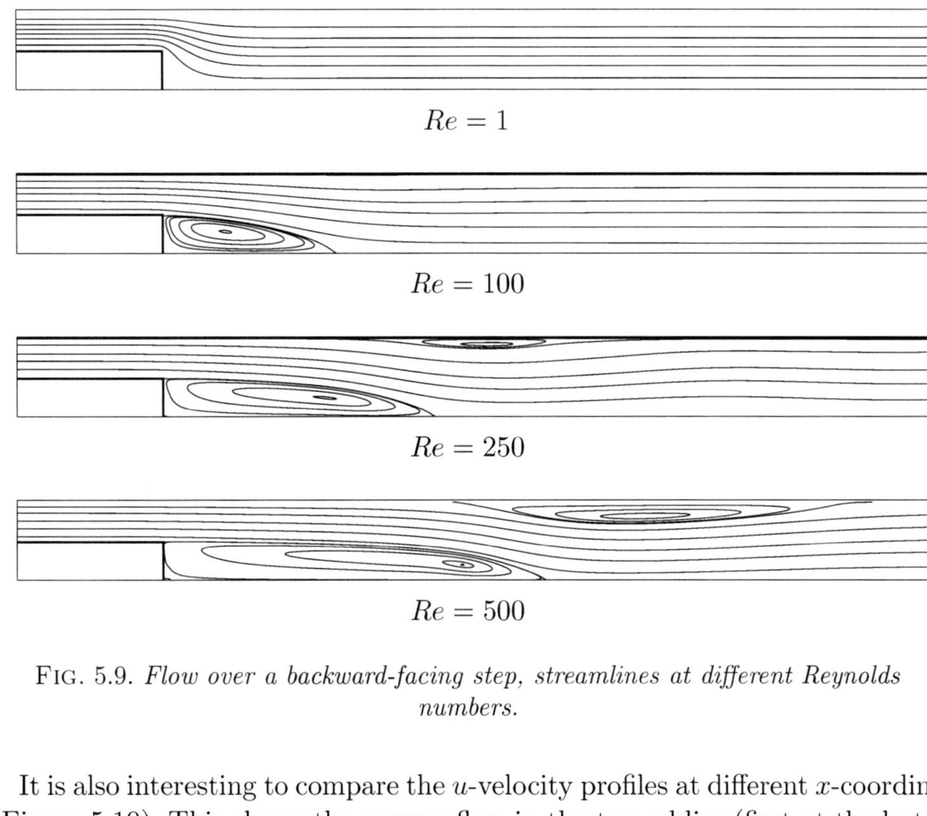

$Re = 1$

$Re = 100$

$Re = 250$

$Re = 500$

FIG. 5.9. *Flow over a backward-facing step, streamlines at different Reynolds numbers.*

It is also interesting to compare the u-velocity profiles at different x-coordinates (Figure 5.10). This shows the reverse flow in the two eddies (first at the bottom, then later at the top). We can also observe how the constant inflow velocity profile widens to cover the complete channel width by the time it reaches the outflow boundary.

5.3 Flow Past an Obstacle

The flow over immersed obstacles is another problem which has been extensively investigated both experimentally and numerically (cf. [Braza et al., 1986], [Dennis

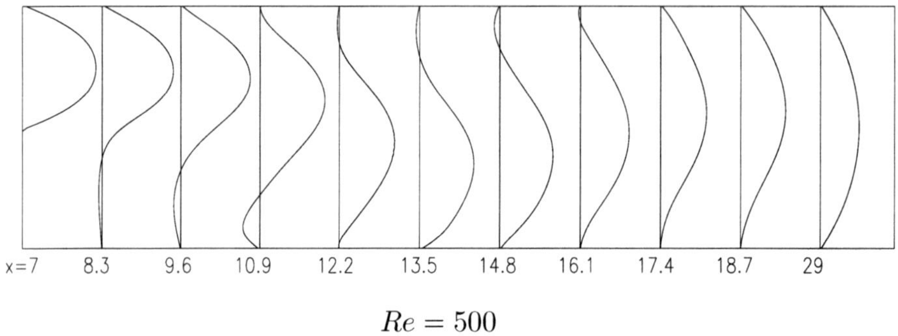

FIG. 5.10. *Flow over a backward-facing step, horizontal velocity profiles.*

& Chang, 1970], [Eaton, 1987], [Perry et al., 1982], [Rannacher, 1992], [Rosenfeld et al., 1991], [Tritton, 1959]). Here we consider the flow past a disk (or cylinder, respectively) immersed in a channel. The inflow velocity at the left boundary has a parabolic profile with a maximal value of $u = 1.5$; the disk measures $d = 1.0$ in diameter and is situated at a distance of 1.5 from the left, .5 from the right, and 1.6 from the upper boundary. For the remaining parameters, we have used

```
imax = 220,      jmax = 41,  xlength = 22,  ylength = 4.1,
delt = 0.0028,   tau = 0.5,  t_end = 20.0,
eps = 0.01,      omg = 1.7,  gamma = 0.9,   itermax = 100,
GX = 0.0,        GY = 0.0,
UI = 1.0,        VI = 0.0,   PI = 0.0,
wW = 3,          wE = 3,     wN = 2,        wS = 2.
```

Physical experiments have shown (see the figures in [Nakayama, 1988, p. 8 ff.]) the structure of the flow to undergo fundamental changes as the Reynolds number increases. For highly viscous fluids ($Re < 4$), the flow divides ahead of the obstacle only to reunite immediately behind it (see Figure 5.11). At lower viscosities ($4 < Re < 40$), the friction along the obstacle surface is no longer strong enough to reunite the two flow segments right away; rather, these separate from the obstacle and re-emerge somewhat later, allowing two steady symmetric eddies to form in the resulting gap (Figures 5.12 and 5.13). In both cases, however, the flow reaches steady state, hence streamlines coincide with streaklines.

When the Reynolds number increases above 40, the flow becomes unsymmetric and unsteady and eddies are shed alternately from the upper and lower edges of the disk in a time-periodic fashion. This trail of vortices in the wake is known as the *Kármán vortex street.*[41] The forces on the obstacle also alternate in direction periodically; the frequency of this period is measured by the *Strouhal number*

[41] At very high Reynolds numbers this flow becomes turbulent (see Chapter 10).

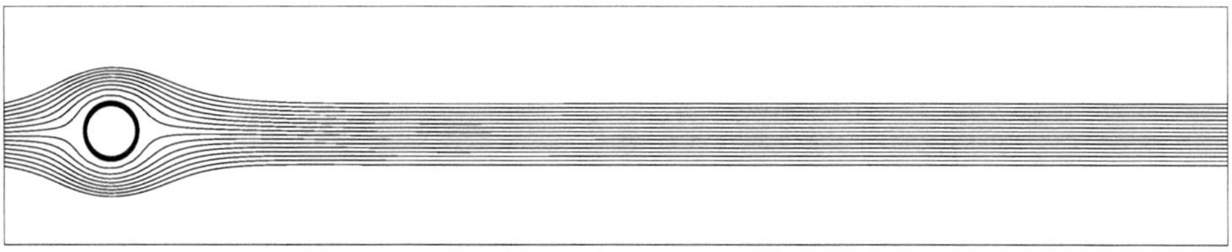

FIG. 5.11. *Flow around a disk ($Re = 1$), streamlines at steady state.*

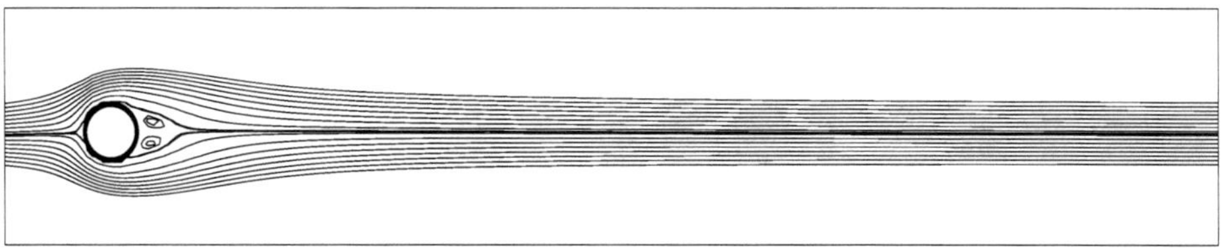

FIG. 5.12. *Flow around a disk ($Re = 20$), streamlines at steady state.*

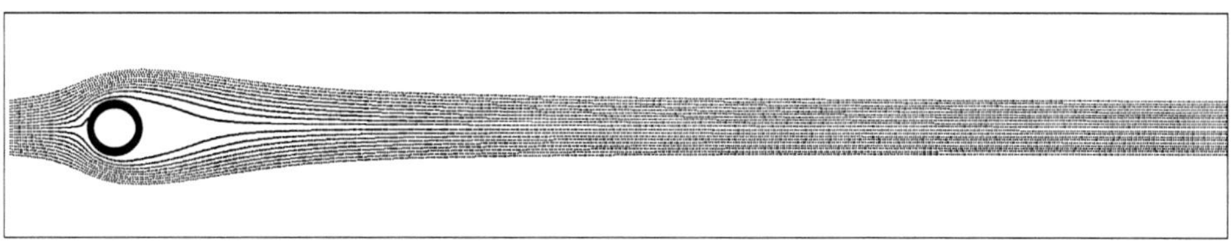

FIG. 5.13. *Flow around a disk ($Re = 20$), streaklines at steady state.*

$$St := d\,f/u_{\text{in}},$$

in which d is the diameter of the disk and u_{in} is the (mean) inflow velocity. The relation $St = 0.212 - 5.35/Re$ between the Strouhal number and the Reynolds number was obtained experimentally in [Hammache & Gharib, 1991]. From our simulations, we were able to determine a period length of $t = 1/f = 0.43$ at $Re = 100$. The numerically determined value of the Strouhal number of 0.157 agrees very well with the experimental value of 0.158. The proportionality of flow velocity and period frequency is utilized in the design of instruments for measuring the flow velocity. For this purpose, a prism is introduced into the flow to generate eddies whose frequency is then measured to infer the velocity.

Because the flow remains unsteady for large Reynolds numbers, streamlines and streaklines will differ. This can be well observed in Figures 5.14 and 5.15,

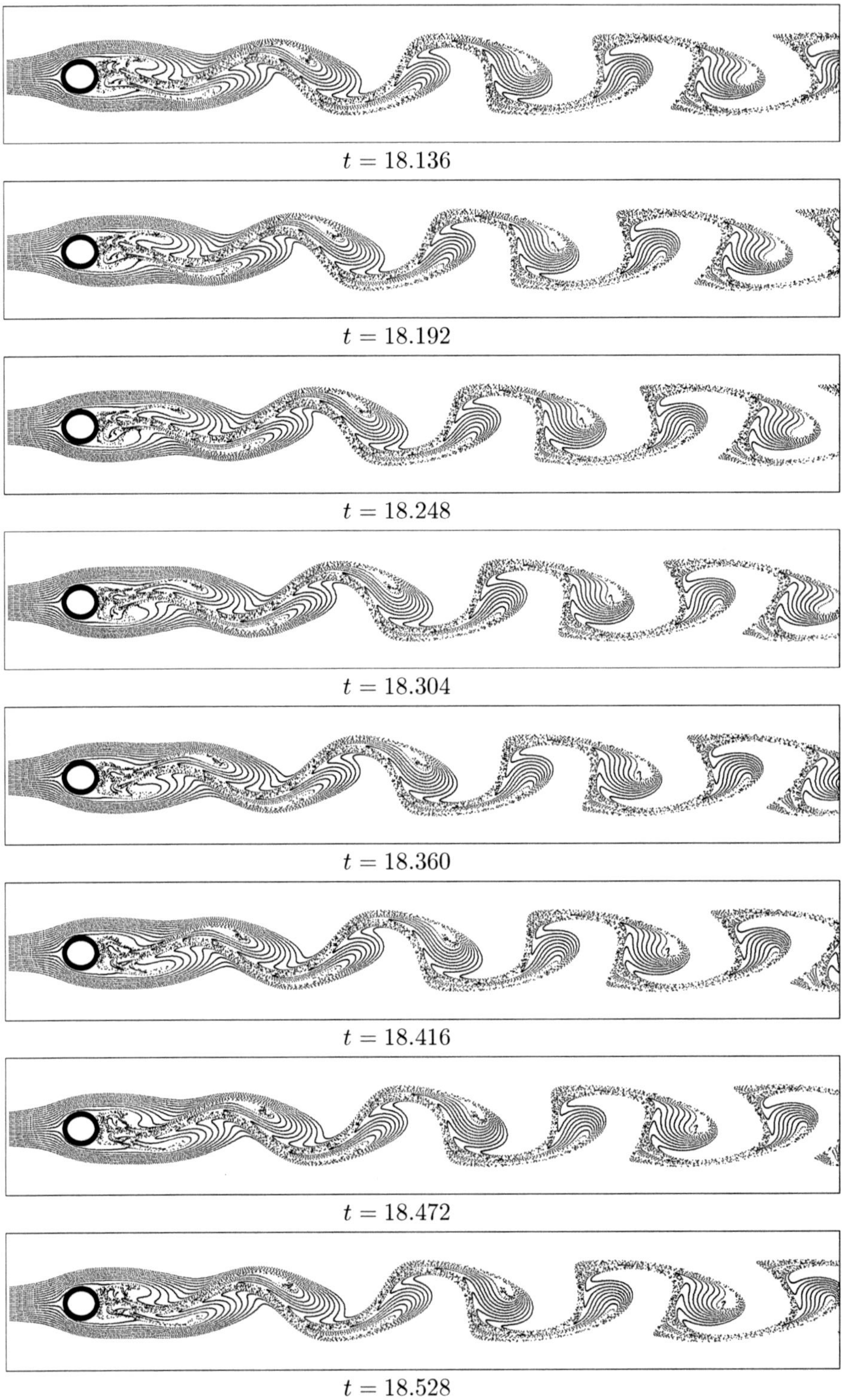

FIG. 5.14. *Flow around a disk ($Re = 100$), streaklines at various time steps.*

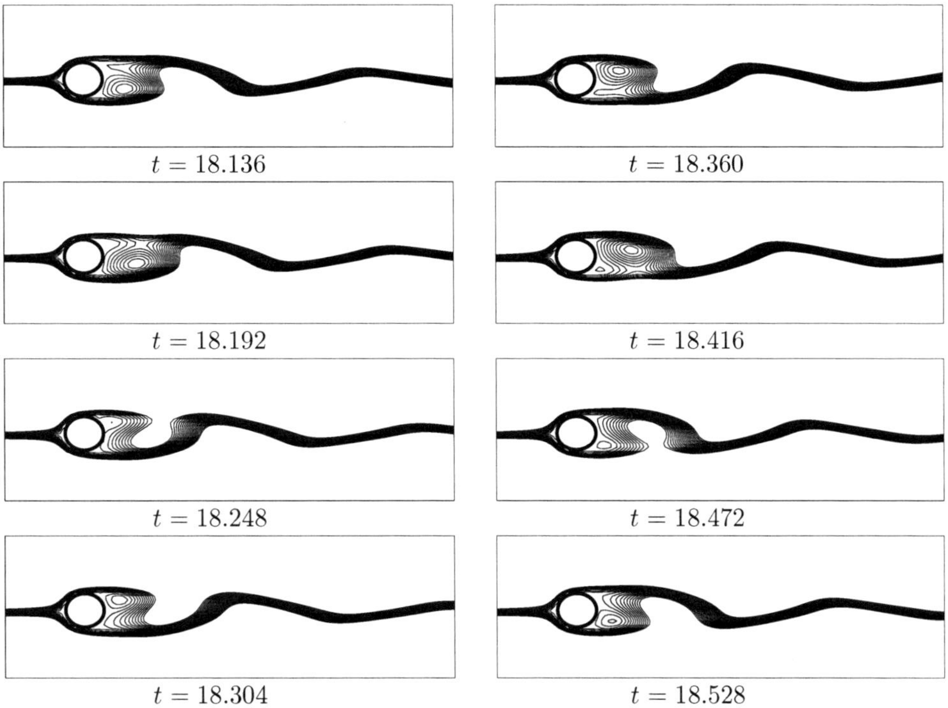

FIG. 5.15. *Flow around a disk ($Re = 100$), streamlines at various time steps.*

each of which displays roughly one period. We remark that Figure 5.15 contains only the left half of the flow domain, and only a small number of the streamlines in the vicinity of the obstacle (cf. also [Kim & Choudhury, 1993]) is displayed.

5.4 Pipe Junction

We continue with some examples which are not standard CFD test cases, beginning with a problem for which the boundary conditions change in time. Motivated by [Prandtl, 1936b], we simulate the flow through a pipe junction as shown in Figure 5.16. The fluid enters through the opening on the left with a constant velocity ($u_{\text{in}} = 1.0$). Initially, the pipe is open on the lower right end and closed at the upper right end. At $t = 20$, we close the bottom and open the top, and at $t = 60$, we return to the initial configuration. In the sequence of images in Figure 5.16, the flow is visualized by particles entering on the left at the start. The obstacle domain in this case is defined as $[0, a - b/3] \times [0, b/3]$ and $[0, a - b/3] \times [2b/3, b]$. The parameters used are

```
imax = 60,     jmax = 60,  xlength = 12,    ylength = 12,
delt = 0.05,   tau = 0.5,  t_end = 100.0,
eps = 0.001,   omg = 1.7,  gamma = 0.7,     itermax = 100,
GX = 0.0,      GY = 0.0,
UI = 1.0,      VI = 0.0,   PI = 0.0,
wW = 3,        wE = 2,     wN = 2 or 3,     wS = 3 or 2.
```

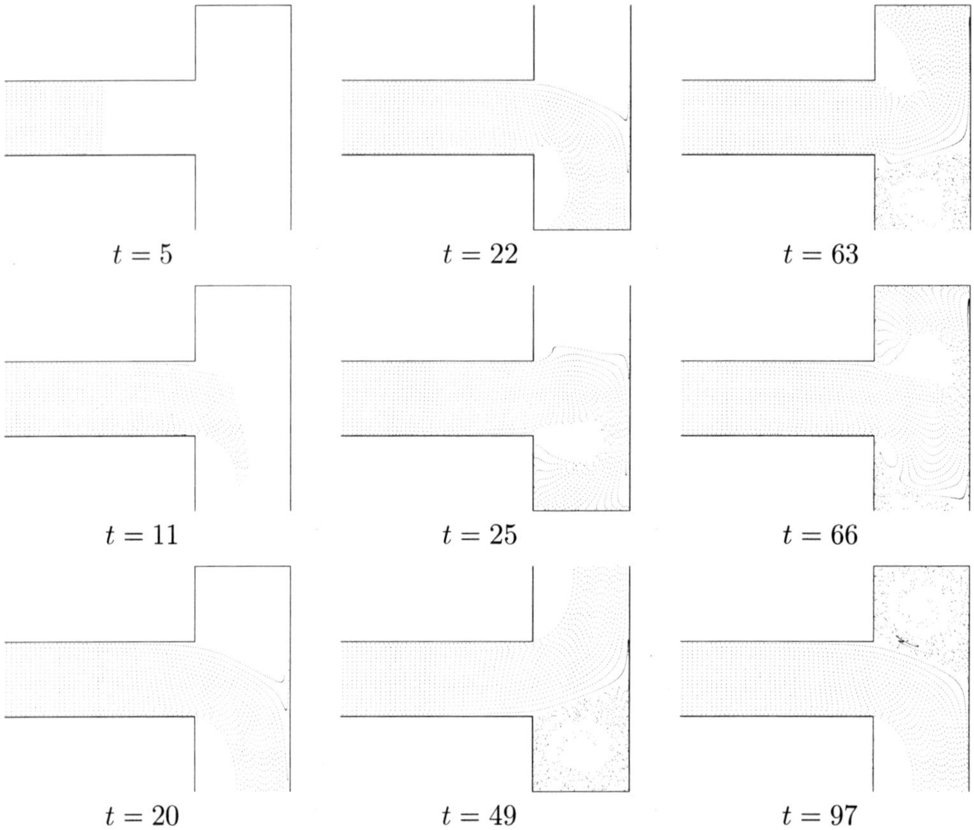

FIG. 5.16. *Flow through a T-shaped junction, time evolution at* $Re = 1000$.

5.5 Flow through Complex Geometries

By employing the flag array, relatively complex and irregular flow geometries may be described. This description becomes more accurate the finer the grid is chosen.[42]

Such an example is given by the flow in porous media. This is an important area of research with numerous applications, among them geophysics (oil production, groundwater transport) and chemical engineering (calculation of flows through filters and catalytic converters). A thorough mathematical account of the simulation of the flow in porous media can be found in [Douglas & Hornung, 1993] and [Russell, 1992], among others.

Usually, the large scale behavior of the fluid is of interest and only the mean

[42]Many flow solvers employ boundary-fitted grids defined by transformations of the domain and thus avoid having to approximate obstacles by vertical and horizontal boundary segments. This, however, requires an analytic description of the boundary, which is very difficult to implement for highly irregular domains.

flow through the porous medium on a macroscopic level is considered. This is described by Darcy's law,

$$\tilde{q} = -\frac{k}{\mu}\,\mathrm{grad}\tilde{p}, \tag{5.1}$$

in which $\tilde{p}$ denotes the averaged pressure, μ the dynamic viscosity, $\tilde{q}$ the average mass flux, and k the permeability of the porous medium. For a given sample, the value of k can be determined in the laboratory. This was first accomplished by the French engineer Henri Darcy in filter experiments during which he discovered the relation (5.1) which bears his name.

To model the permeability, it may be interesting to simulate this experiment numerically: the flow through a small sample of the porous medium under consideration is simulated at the microscopic level by solving the Navier–Stokes equations in the geometrically complex pore space of the sample. By averaging the steady state solution, the permeability of the sample may be computed using (5.1) (cf. [Knapek, 1994]).

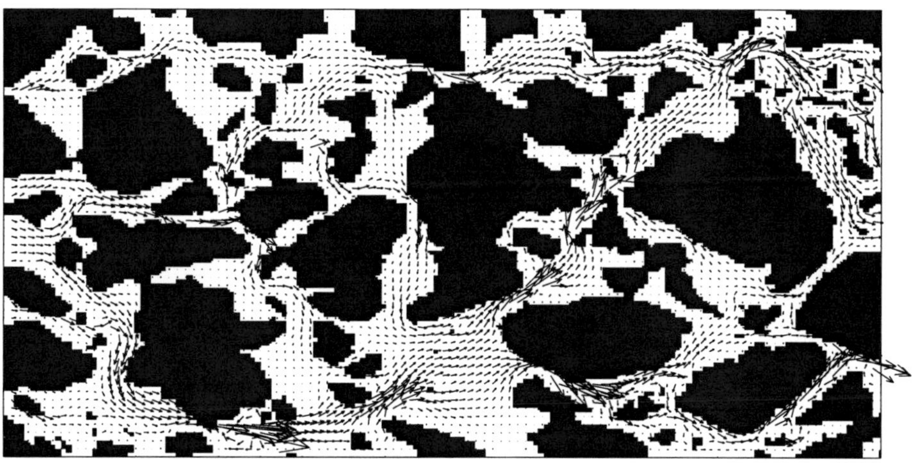

FIG. 5.17. *Flow through a porous medium, geometric configuration and velocity vectors.*

Motivated by [Bernsdorf, 1994], we have scanned the electron microscope image of a sediment sample and used the black-and-white pixel image to define the flag array. Using the inflow velocity $u_{\text{in}} = 1.0$ at the left boundary along with the parameters

```
imax = 422,    jmax = 216,   xlength = 84.4,  ylength = 43.2,
delt = 0.01,   tau = 0.5,    t_end = 15.0,
eps = 0.01,    omg = 1.7,    gamma = 0.9,     itermax = 200,
GX = 0.0,      GY = 0.0,     Re = 10,
UI = 1.0,      VI = 0.0,     PI = 0.0,
wW = 3,        wE = 3,       wN = 2,          wS = 2,
```

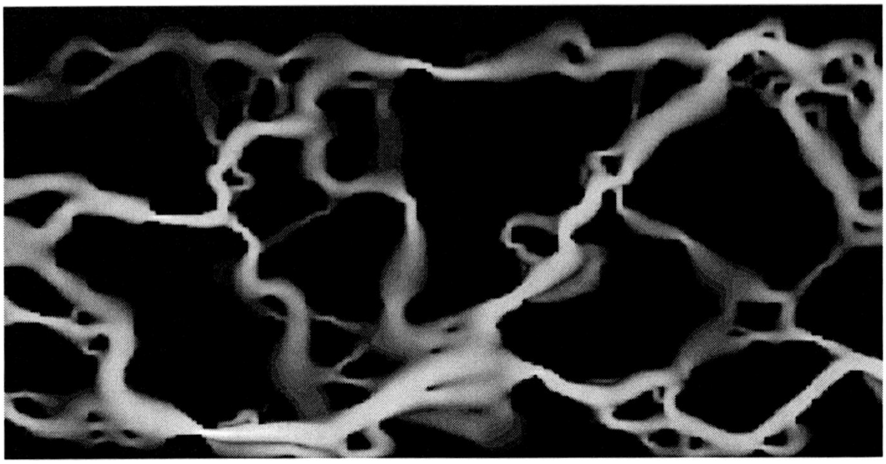

FIG. 5.18. *Flow through a porous medium, grayscale image of velocity magnitude.*

we obtain the steady state results depicted in Figures 5.17 and 5.18.

5.6 Fluid-Structure Interaction

Another current area of research is the study of the interactions between fluids and solid bodies. This entails investigating the effect of moving bodies on fluids in motion as well as the study of how a moving fluid can deform an elastic body or transport movable bodies.

We consider a micropump as it is used, e.g., in medical engineering. Its small dimensions lead to a small Reynolds number, and hence the flow generally remains laminar and does not become turbulent. A piston is moved back and forth at the right boundary, which in our case is implemented simply by imposing different inflow conditions ($u_{in} = 1.0$ and $u_{in} = -1.0$, resp.). At the left boundary, the normal derivatives of u and v are set to zero, implementing an outflow boundary condition. When the piston is moved to the right, a drop in pressure results in the chamber, causing the upper valve to open and the lower valve to close. This allows the fluid to enter the chamber through the upper channel (see Figure 5.19, left).

Conversely, when the piston is moved in the opposite direction, the upper valve blocks while the lower valve opens, permitting the fluid to leave via the lower channel (see Figure 5.19, right). The valves themselves move as a result of the pressure in the flow according to the laws of elasticity. In our simulation, the valve movement is simulated by moving the boundary cells [Bungartz & Schulte, 1995], [Meier, 1995]. The parameters used here are

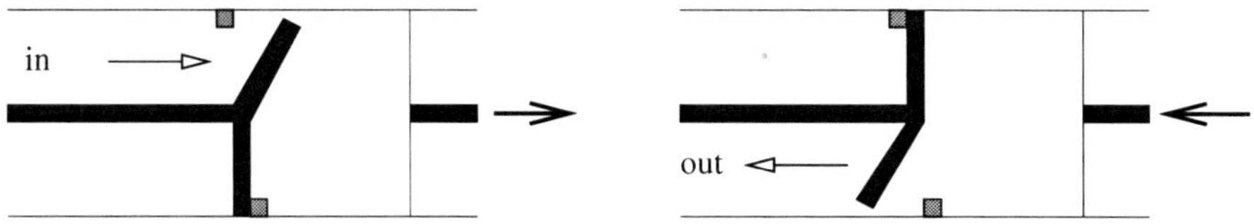

FIG. 5.19. *Micropump, schematic representation.*

```
imax = 128,    jmax = 65,   xlength = 20.48,   ylength = 10.4,
delt = 0.02,   tau = 0.5,
eps = 0.001,   omg = 1.7,   gamma = 0.9,       itermax = 200,
GX = 0.0,      GY = 0.0,    Re = 10,
UI = 1.0,      VI = 0.0,    PI = 0.0,
wW = 3,        wE = 3,      wN = 2,            wS = 2.
```

In Figure 5.20, the results of simulating one cycle are shown by way of the flow of injected particles. In the uppermost image, the piston is moving to the left. The image below shows the situation shortly after the piston has changed direction, when the upper valve is in the process of opening, and the lower valve is in the process of closing. In the third image the lower valve has closed, and the bottom image shows the situation just after the next change in the piston's direction.

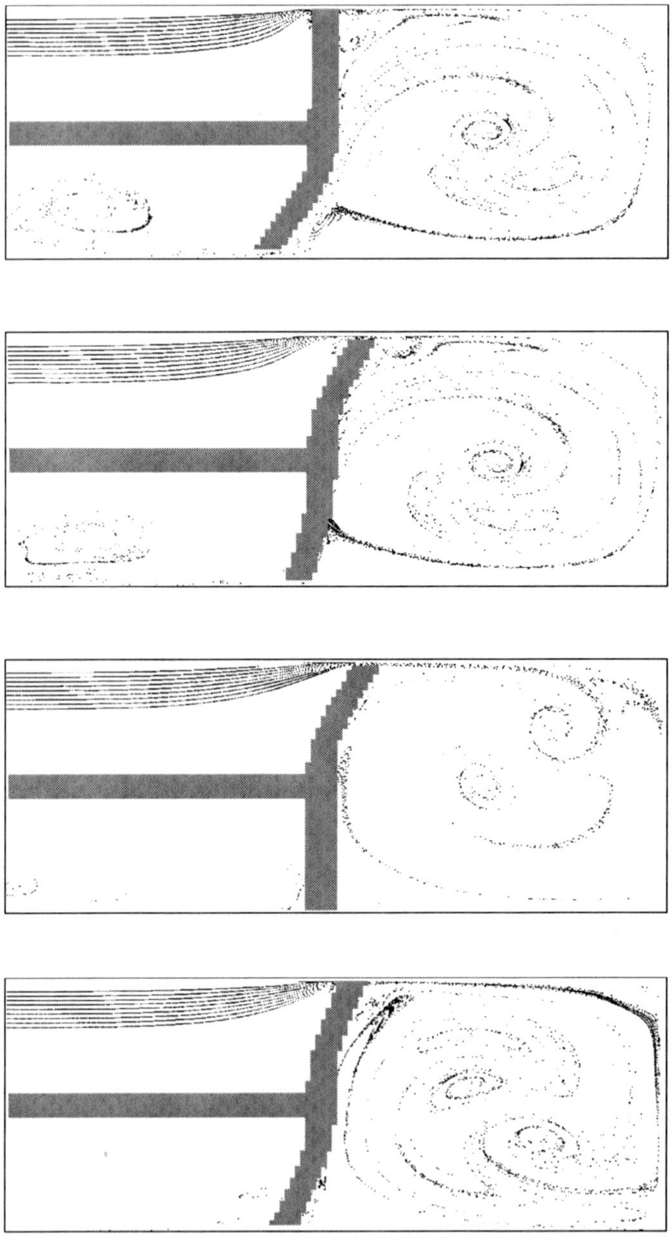

FIG. 5.20. *Micropump, simulation of one cycle.*

Chapter 6

Free Boundary Value Problems

While the flows we have considered so far were all confined to a fixed given domain Ω, we now turn to what are known as *free boundary value problems.* These are problems in which the shape of the domain occupied by the fluid may change with time. This means that, in addition to the pressure p and the velocities u and v, the domain Ω_t, which the fluid occupies at time t, is also to be determined. Besides the Navier–Stokes equations and the initial and boundary conditions at the fixed boundaries, the problem description now also requires the initial configuration Ω_0 of the fluid domain as well as conditions which hold along the free boundary.

The range of applications is very broad. In nature, the flow of a wild brook, the waves on the ocean, and the falling of a raindrop into a puddle are examples of free boundary value problems. In engineering applications, the injection molding of plastics as well as melting, solidification, and coating processes may all be formulated as free boundary value problems. Two-phase and multiphase flows, i.e., the flow of two or more different fluids the interface between which represents a free boundary, also fall into this problem class.

In addition to the algorithm described in Chapter 3 for the approximate solution of the Navier–Stokes equations, the current domain must now be calculated in each time step before solving for the pressure and velocities on this domain. This can be done, for example, using particles which are advanced according to the current velocities from time step to time step and which describe the current state of the fluid. With the help of these particles, the rectangular base domain $\mathcal{G}$ is divided into not just the fluid domain Ω and the obstacle domain $\mathcal{H}$ as in Section 3.4, but rather into the obstacle domain $\mathcal{H}$, the time-dependent fluid domain Ω_t, and the remaining empty domain $\mathcal{G} \setminus \{\mathcal{H} \cup \Omega_t\}$. The interface between the fluid domain and the empty domain then constitutes the free boundary. In Figure 6.1 this is illustrated for a simple injection molding example. It shows the molding of a rectangular plastic object in which the liquid plastic mass is injected through an opening in the left boundary, eventually filling the domain $\mathcal{G} \setminus \mathcal{H}$. This, however, takes into account neither the cooling process nor the fact that the outflow of displaced air should also be simulated.

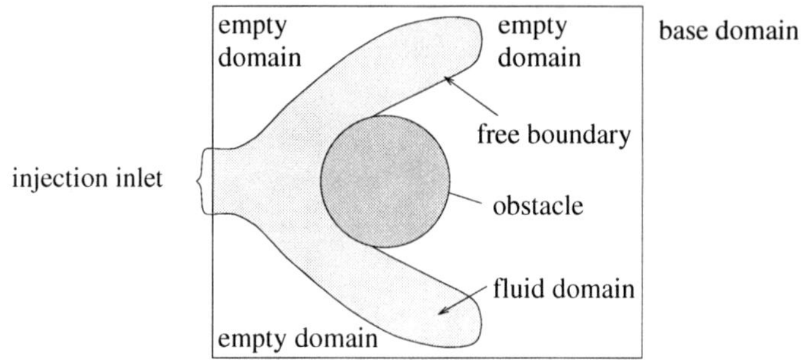

FIG. 6.1. *Injection molding of a rectangular plastic part containing a round hole.*

6.1 Determination of the Domain Shape

To determine the domain to be treated in a given time step we use particles similar to those used for visualization in Section 4.2.

At time $t = 0$, the initial geometry of the domain Ω_0 occupied by the fluid must first be specified. This is done by assigning to each cell of the grid a fixed number of potential particle positions (usually 9 or 16) spaced a constant distance apart. However, only those particles whose potential positions lie inside the initial domain are placed at their positions (see Figure 6.2).

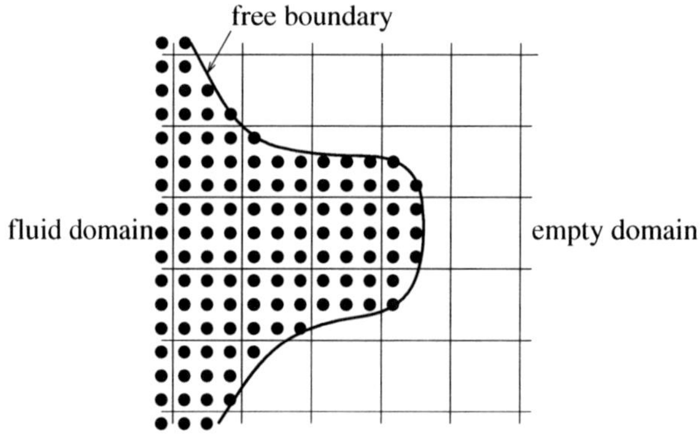

FIG. 6.2. *Particles in an initial configuration with nine potential particle positions per cell.*

Now assume that, at time step n, the particle positions at time t_n are known. These current positions may be used to determine whether a cell is empty, is located entirely within the fluid domain, or lies on the boundary of the latter. Then the corresponding boundary values can be determined along the free boundary before computing the pressure and velocity values in the interior of the fluid domain at time t_{n+1}. The values outside the fluid domain are of no interest. Finally, at the end of the time step, the particle positions at time t_{n+1} are updated using the newly calculated velocities $u^{(n+1)}$ and $v^{(n+1)}$.

Using the particle positions, we now partition the cells of our staggered grid into different types. Besides the obstacle cells we defined in Section 3.4, we now distinguish

- empty cells: cells containing no particles (or fluid, resp.),
- interior cells: cells that contain particles and do not border on *empty cells*,
- surface cells: cells that contain particles and do border on *empty cells*.

Thus, the fluid domain is represented by the *interior cells* and the free boundary is described approximately by the *surface cells*.

The partitioning of the cells may be accomplished by extending the integer array `int **FLAG`, previously introduced in Section 3.4. In addition to the values `C_B` and `B_N`, ..., `B_NW`[43] for the obstacle cells approximating the obstacle domain $\mathcal{H}$, let the macro `C_F` denote the *interior cells* and let `C_X` denote the *empty cells*.

Like the set of obstacle boundary cells, the set of surface cells must also be further subdivided. We do this by specifying in these cells' flag array entries those directions in which empty cells are found. Thus, as an example, a cell whose only empty neighbor lies to its left may be coded as `C_W`, those whose lower and right neighbors are empty, with `C_SE`, and one with all four neighbors empty, with `C_NSWE`. This results in fifteen different types of surface cells to be distinguished. The values `C_xy` are again represented by different but otherwise arbitrary integer constants.

The marking of the cells begins with a loop over all particles, in which all cells containing particles are marked with `C_F`. Those remaining cells which are not obstacle cells are given the flag `C_X`. At the same time, those particles which have left the base domain $\mathcal{G}$ or which have ventured into obstacle cells, are eliminated.[44]

Those cells which have received a `C_F` flag in the first loop must then be further subdivided into true interior cells and surface cells by determining which of the four neighboring cells, if any, are empty, and then assigning the corresponding surface cell flag where appropriate.

Aside from this technique of determining the domain shape by particles distributed over the entire fluid domain, there are also other possibilities requiring much less computation and storage. These, however, either have other disadvantages or are more complex to describe and implement.

The free boundary could, for example, also be described by the graph of an elevation function $h = h(x)$ or $h = h(y)$ as suggested in [Hirt et al., 1975]. Naturally, this works only when the free boundary can be expressed as a function of x or y. This method can no longer be used once the boundary intersects a y- or x-coordinate line more than once (liquid drop, breaking wave, injection molding).

In [Nichols & Hirt, 1971] as well as [Chen et al., 1991], particles are used only for the description of the free boundary and are also advanced as in the method we

[43] In contrast to the definition of boundary cells in Section 3.4, we here regard as boundary cells all those obstacle cells lying adjacent to interior, empty, or surface cells.

[44] Cf. also footnote 37 on page 58.

have described. This eliminates the restriction on the shape of the free boundary mentioned above. Problems may arise, however, when two free boundaries collide. Furthermore, the extension to three dimensions is not quite simple.

In the *volume-of-fluid* (VOF) method [Hirt & Nichols, 1981], which was extended to the treatment of two-phase flows with surface tension by Lafaurie et al. in the *SURFER* code [Lafaurie et al., 1994], the volume fraction of the cell volume of each cell filled with fluid is stored in a separate array. A value of one characterizes interior cells, empty cells have the value zero, and those of the surface cells vary between zero and one. If the direction of the free boundary in the surface cells is also determined, then the change of volume fraction may be computed using the current velocities.

6.2 Conditions along the Free Boundary

The Mathematical Model

We begin by briefly considering the physical conditions which hold at the interface separating *two* different fluids forming an a priori unknown free boundary Γ_f. Along this free boundary, the force acting on the interface is in the normal direction and its magnitude is proportional to the interface's mean curvature. Using the stress tensor

$$\boldsymbol{\sigma} = (-p + \lambda\, \mathrm{div}\vec{u})\mathbf{I} + 2\mu\boldsymbol{\delta}$$

from (2.12) (page 17), this may be expressed as

$$(\boldsymbol{\sigma}^{(1)} - \boldsymbol{\sigma}^{(2)})\,\vec{n} = k\left(\frac{1}{R_1} + \frac{1}{R_2}\right)\vec{n} \quad \text{on} \quad \Gamma_f \tag{6.1}$$

(cf. [Landau, 1959]), where $\boldsymbol{\sigma}^{(1)}$ denotes the stress tensor in one fluid and $\boldsymbol{\sigma}^{(2)}$ that in the other. Furthermore, R_1 and R_2 are the interface's principal radii of curvature (cf., e.g., [Berger & Gastiaux, 1988]) and k is the coefficient of surface tension, a material constant depending only on the two fluids.

In what follows, we will consider only *one* fluid in a vacuum and in two dimensions, so that we can simplify matters by setting $\boldsymbol{\sigma}^{(2)} = \mathbf{0}$ ($Re^{(2)} = \infty$, $p^{(2)}$ normalized to zero), $R_2 = \infty$, and $1/R_1 =: \kappa$, the curvature of the free boundary.

Using the continuity equation $\mathrm{div}\vec{u} = 0$ and the normalization with respect to the density ϱ_0 as in (2.16) (page 17), the two-dimensional stress tensor is given by

$$\boldsymbol{\sigma}^{(1)} = -\begin{pmatrix} p & 0 \\ 0 & p \end{pmatrix} + \frac{1}{Re}\begin{pmatrix} 2\,\partial u/\partial x & \partial u/\partial y + \partial v/\partial x \\ \partial u/\partial y + \partial v/\partial x & 2\,\partial v/\partial y \end{pmatrix}.$$

If we now multiply (6.1) once with the unit vector orthogonal to the interface, i.e., the normal vector $\vec{n} := (n_x, n_y)^T$, and once with the unit vector parallel to the interface, i.e., the tangent vector $\vec{m} := (m_x, m_y)^T$, we obtain the two boundary conditions

$$-p + \frac{2}{Re}\left(n_x\, n_x\, \frac{\partial u}{\partial x} + n_x\, n_y \left(\frac{\partial u}{\partial y} + \frac{\partial v}{\partial x}\right) + n_y\, n_y\, \frac{\partial v}{\partial y}\right) = k\,\kappa \tag{6.2}$$

and

$$2\, n_x\, m_x\, \frac{\partial u}{\partial x} + (n_x\, m_y + n_y\, m_x)\left(\frac{\partial u}{\partial y} + \frac{\partial v}{\partial x}\right) + 2\, n_y\, m_y\, \frac{\partial v}{\partial y} = 0. \tag{6.3}$$

These boundary conditions must now be suitably discretized at the centers of the surface cells.[45] For simplicity, we restrict ourselves to the case of vanishing surface tension ($k = 0$).[46]

Discretization of the Conditions along the Free Boundary

We must first address the question of at which grid points we require boundary values for u, v, and p for calculating new pressure and velocity values in the fluid interior.

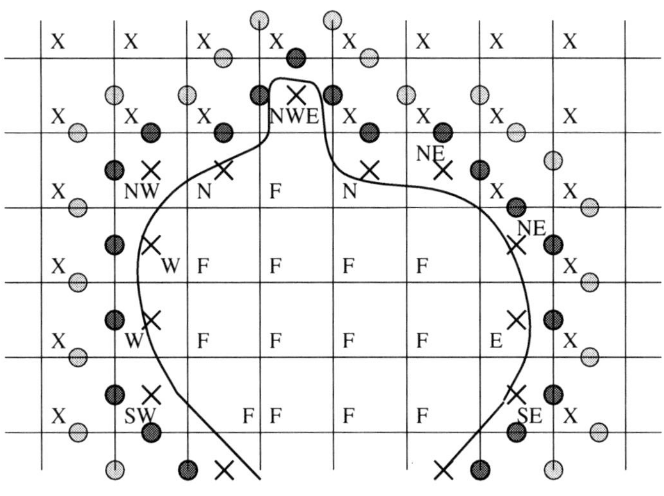

FIG. 6.3. *Required boundary data.*

These boundary values are as follows (see Figure 6.3):

- pressure in the surface cells (crosses),
- velocities at edges separating surface cells from empty cells (dark circles),
- velocities at edges separating two empty cells adjacent to a corner of a surface cell (light circles).

[45]In [Nichols & Hirt, 1971] a method is described which discretizes at points on the free boundary (which is determined with greater accuracy) rather than at the surface cell centers.

[46]At greatly reduced gravity levels (microgravitation) or for very thin fluid layers, surface tension may have a very strong effect on the flow. For the treatment of surface tension, cf., e.g., [Daly & Pracht, 1968], [Lafaurie et al., 1994].

With this boundary data available, the pressure in the interior cells may be calculated using the F and G values on adjacent edges according to (3.38) on page 35. Using the pressure values thus computed in interior and surface cells, the velocities on the edges not bordering on empty cells can be computed according to (3.34) and (3.35) (page 34) using appropriate F and G values. The new particle positions are then determined following (4.3) on page 55.

Since the above-mentioned pressure and velocity values are calculated from neighboring velocity values (see below) and since all empty cells adjacent to surface cells could become surface cells in the next time step, we must ensure that "sensible" velocity values already exist at the points marked with light circles in Figure 6.3. This clearly illustrates the necessity of the CFL conditions in (3.49) (page 39), as these ensure that particles travel a distance of at most one cell per time step.

In discretizing the conditions (6.2) and (6.3) at the free boundary, we must distinguish five types of surface cells.

1. *Cells with one empty neighbor:* We assume the free boundary to lie almost parallel to the grid lines, resulting in either of n_y and m_x or n_x and m_y being very small. Equation (6.2) (with zero right-hand side) may then be approximated by

$$p = \frac{2}{Re}\frac{\partial u}{\partial x} \qquad \text{or} \qquad p = \frac{2}{Re}\frac{\partial v}{\partial y}, \tag{6.4}$$

respectively—the first equation pertaining to vertical, the second to horizontal boundaries. Equation (6.3) becomes

$$\frac{\partial u}{\partial y} + \frac{\partial v}{\partial x} = 0\,. \tag{6.5}$$

The discretization of these formulas is explained using a `C_E`-cell as an example (see Figure 6.4). The pressure p in cell (i,j) is determined from equation (6.4):

$$p_{i,j} = \frac{2}{Re}\frac{(u_{i,j} - u_{i-1,j})}{\delta x}\,. \tag{6.6}$$

Prior to this, the velocity value $u_{i,j}$ on the edge between the surface cell and the empty cell is chosen so that the discrete counterpart of the continuity equation $\partial u/\partial x + \partial v/\partial y = 0$ is satisfied in cell (i,j):

$$u_{i,j} = u_{i-1,j} - \frac{\delta x}{\delta y}(v_{i,j} - v_{i,j-1})\,. \tag{6.7}$$

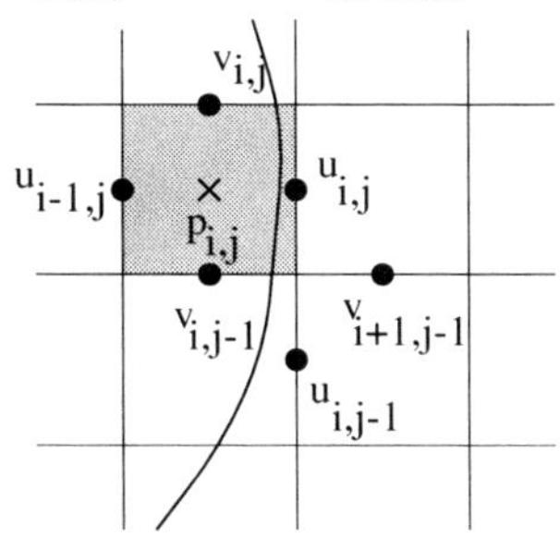

FIG. 6.4.
Cell with flag C_E.

For the calculation of the velocity $v_{i+1,j-1}$ belonging to an edge separating two empty cells and which meets a surface cell, we make use of condition (6.5):

$$v_{i+1,j-1} = v_{i,j-1} - \frac{\delta x}{\delta y}(u_{i,j} - u_{i,j-1})\,. \tag{6.8}$$

However, this may be done only if cell $(i+1,j-1)$ is empty, since otherwise $v_{i+1,j-1}$ is determined from the continuity equation in cell $(i+1,j-1)$. Since the values $u_{i,j}$ and $u_{i,j-1}$ are required for the evaluation of (6.8), the latter must occur after the computation of $u_{i,j}$ according to (6.7); $u_{i,j-1}$ is already known if the computation proceeds from the lower left to the upper right $(i = 1, \ldots, i_{\max},\ j = 1, \ldots, j_{\max})$.

For the remaining cells with *one* empty neighbor (flags C_N, C_W, C_S), formulas analogous to these are applied. Note that, in some cases, u and v, δx and δy as well as the signs are switched. Formula (6.5) is applied to each of the left and lower adjoining edges, respectively, to have all required values available (see also Figure 6.8).

2. *Cells with two neighboring empty cells sharing a common corner:* In this case the free boundary can be approximated by a line segment intersecting the grid lines at an angle of 45° such that n_x, n_y, m_x, and m_y are all of equal magnitude and differ only in sign.

   Formula (6.2) then reduces to

$$p = \pm\frac{1}{Re}\left(\frac{\partial u}{\partial y} + \frac{\partial v}{\partial x}\right), \tag{6.9}$$

   in which the plus sign applies to cells with flags C_NE or C_SW, and the minus sign to cells with flags C_NW and C_SE. Formula (6.3) becomes

$$\frac{\partial u}{\partial x} - \frac{\partial v}{\partial y} = 0\,. \tag{6.10}$$

   The discretizations are exemplified here using a C_SE-cell (see Figure 6.5). Equation (6.10) and the continuity equation for cell (i,j) yield $\partial u/\partial x = 0$ and $\partial v/\partial y = 0$. In discretized form, this is

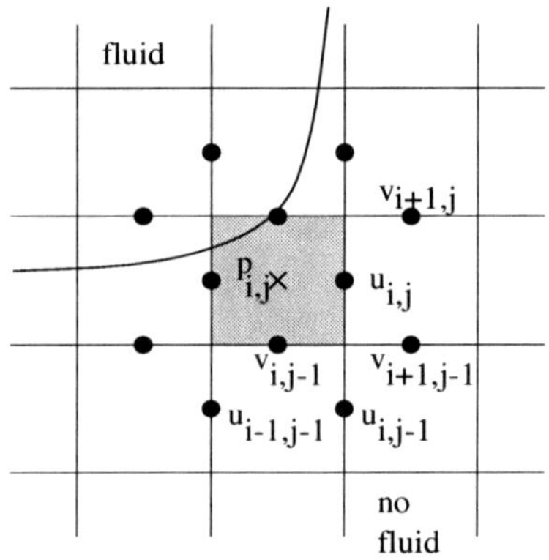

FIG. 6.5. *Cell with flag* C_SE.

$$u_{i,j} = u_{i-1,j}\,, \qquad v_{i,j-1} = v_{i,j}\,, \tag{6.11}$$

whereas the pressure in cell (i,j) may be determined from equation (6.9) as

$$p_{i,j} = -\frac{1}{2\,Re}\left(\frac{u_{i,j+1} + u_{i-1,j+1} - u_{i,j} - u_{i-1,j}}{\delta y} + \frac{v_{i,j} + v_{i,j-1} - v_{i-1,j} - v_{i-1,j-1}}{\delta x}\right). \tag{6.12}$$

When either of cells $(i-1, j-1)$ or $(i+1, j+1)$ is empty, then the free boundary intersects either the left or upper edge, respectively, at an angle of nearly 90°. Hence we may use formula (6.5) to approximate $u_{i-1,j-1}$ and $v_{i+1,j}$:

$$\begin{aligned} u_{i-1,j-1} &= u_{i-1,j} + \frac{\delta y}{\delta x}(v_{i,j-1} - v_{i-1,j-1})\,, \\ v_{i+1,j} &= v_{i,j} - \frac{\delta x}{\delta y}(u_{i,j+1} - u_{i,j})\,. \end{aligned} \tag{6.13}$$

Here $v_{i+1,j}$ can only be computed once $u_{i,j+1}$ is known; i.e., $v_{i+1,j}$ is not determined until cell $(i, j+1)$ has been treated. If cells $(i-1, j-1)$ or $(i+1, j+1)$ are not empty, then $u_{i-1,j-1}$ and $v_{i+1,j}$, respectively, are determined when these cells are treated.

The values of $u_{i,j-1}$ and $v_{i+1,j-1}$ at the corner are each set equal to the closest value:

$$u_{i,j-1} = u_{i,j}\,, \qquad v_{i+1,j-1} = v_{i,j-1}\,. \tag{6.14}$$

The three remaining cases with flags C_NW, C_NE, and C_SW are treated analogously. Again, signs as well as δx and δy and indices are exchanged accordingly. Care must also again be taken that the two values, for whose discretization formula (6.5) is employed, may be set only if they lie on an edge level with either the left or lower boundary of cell (i,j) (see Figure 6.8).

In all cases, it must also be noted that the corresponding values may not lie on edges of another surface cell.

3. *Cells with two opposite empty neighbors:* In this as well as the other remaining cases, the boundary conditions can no longer be properly satisfied since the normal and tangent vectors can no longer be uniquely approximated. While these cases would not arise in a sufficiently fine discretization, they must nevertheless be treated in some appropriate manner, since otherwise a new discretization could become necessary at each time step, possibly leading to too many unknowns.

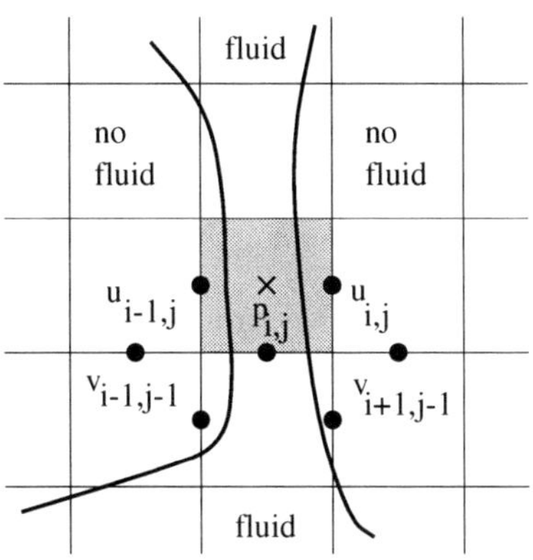

FIG. 6.6. *Cell with flag* C_WE.

We assume here that the fluid in such cells moves only as a result of the body force g_x in the x-direction and g_y in the y-direction, respectively. For cells (i,j) carrying the flag C_WE (see Figure 6.6),

$$u_{i,j}^{\text{new}} = u_{i,j}^{\text{old}} + \delta t\, g_x, \qquad u_{i-1,j}^{\text{new}} = u_{i-1,j}^{\text{old}} + \delta t\, g_x, \tag{6.15}$$

and for cells carrying the flag C_NS,

$$v_{i,j}^{\text{new}} = v_{i,j}^{\text{old}} + \delta t\, g_y, \qquad v_{i,j-1}^{\text{new}} = v_{i,j-1}^{\text{old}} + \delta t\, g_y. \tag{6.16}$$

The velocities $v_{i-1,j-1}$ and $v_{i+1,j-1}$ are computed from (6.5) when cells $(i-1,j-1)$ or $(i+1,j-1)$, respectively, are empty. The pressure in the cell is set equal to that in the surrounding empty cells, which is normalized to zero; i.e., $p_{i,j} = 0$.

4. *Cells with three neighboring empty cells:* In this case, the boundary conditions can again no longer be satisfied exactly. The necessary steps are demonstrated using as an example a cell with flag C_NWE (see Figure 6.7).

   The velocities $u_{i-1,j}$ and $u_{i,j}$ on opposite edges, each bordering on empty neighbors, are treated as in (6.15). The velocity $v_{i,j}$ on the third cell next to an empty neighbor is set to satisfy the continuity equation in cell (i,j).

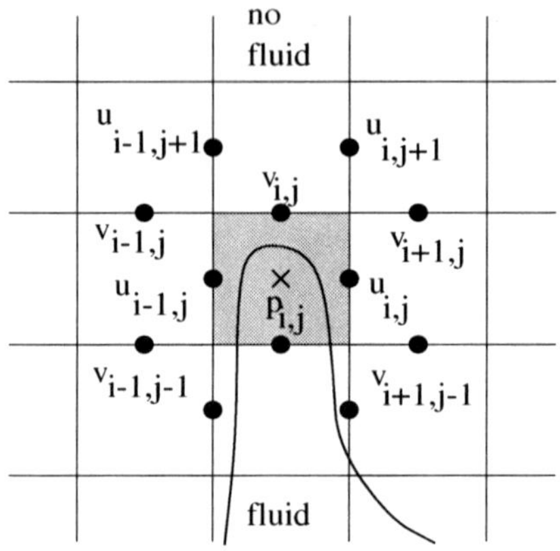

FIG. 6.7. *Cell with flag* C_NWE.

The velocities $u_{i-1,j+1}$, $u_{i,j+1}$, $v_{i-1,j}$, and $v_{i+1,j}$ adjacent to the corner are set equal to whichever velocity on an edge of cell (i,j) (among $u_{i-1,j}$, $u_{i,j}$ and $v_{i,j}$) lies closest—assuming cells $(i-1,j+1)$ and $(i+1,j+1)$, respectively, are empty—just as in the case of two neighboring empty cells sharing a corner. Furthermore, the velocities $v_{i-1,j-1}$ and $v_{i+1,j-1}$ from (6.5) must still be determined when cells $(i-1,j-1)$ and $(i+1,j-1)$, respectively, are empty.

The pressure $p_{i,j}$ is set to zero as in cells with two opposite empty neighbors. Cells with flags C_NSE, C_NSW, and C_SWE are treated analogously, except that the velocities $u_{i-1,j+1}$ and $u_{i-1,j-1}$ corresponding to $v_{i-1,j-1}$ and $v_{i+1,j-1}$ are only computed in the C_NSE case, since only in this case are all required values readily available (compare Figure 6.8).

5. *Cells with four empty neighbors:* In this case, we compute all velocities at the edges of the cells marked with C_NSWE according to (6.15) and (6.16). The eight neighboring values are set equal to the closest values on edges of the surface cell whenever these border on empty cells (e.g., $u_{i,j+1} = u_{i,j}$ or $v_{i-1,j} = v_{i,j}$). The pressure is again set to zero.

Figure 6.8 illustrates once more which of the velocity values marked with light circles in Figure 6.3 still need to be calculated in the course of processing the corresponding surface cells; this assumes that the cells are traversed from the lower left to the upper right in setting the boundary values. The values computed via the discrete version of equation (6.5) are marked with a black dot; those set analogously to (6.14) are marked with a white dot. Each of these formulas, however, is applied only if the adjacent gray cells are empty, since otherwise the velocities are determined when processing these cells. The abbreviation "S/F" in Figure 6.8 denotes cells lying either in the fluid interior or on the free boundary.

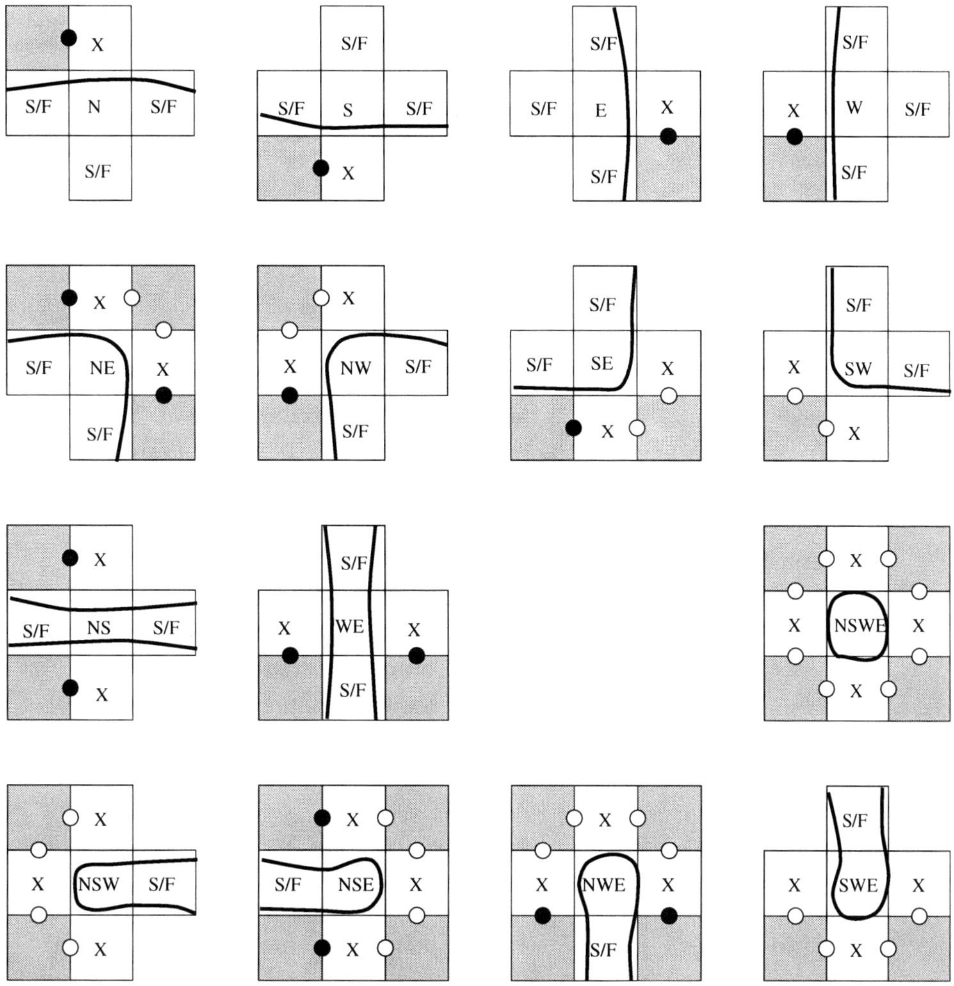

FIG. 6.8. *Surface cells.*

6.3 The Extended Algorithm

Let us now summarize the algorithm extended for the treatment of free boundaries. We emphasize once again that, before advancing the particles during each time step, the boundary values—both free and fixed—should be set once again in order to have all velocities available at time t_{n+1}; otherwise one would use velocities belonging to time t_{n+1} in the interior versus velocities belonging to time t_n along the boundaries for the calculation of the new particle positions. It suffices to set the values at the fixed boundary once *before* entering the time-stepping loop as well as within the loop *following* the computation of the interior velocities, whereas the boundary values along the free boundary must be set *twice*: once *after* calculating the new velocities, but also a second time *after* updating the domain shape, which can change due to the advanced particle positions. In summary, we obtain Algorithm 2.

```
Set t := 0, n := 0
Set up the particles in the initial domain Ω_0
Assign initial values to u, v, p
Set the values for u and v along the fixed boundary of Ω_0
While t < t_end
    Choose δt (according to (3.50) if stepsize control is used)
    Mark interior cells and surface cells of Ω_{t_n}
    Determine values of u, v, and p at the free boundary of Ω_{t_n}
    Compute F^(n) and G^(n) inside the fluid domain according to (3.36),(3.37)
    Compute right-hand side of the pressure equation (3.38) for interior cells
    Set it := 0
    While it < it_max and ||r^it|| > eps (or ||r^it|| > eps ||p^0||), resp.
        Perform SOR cycle according to (3.44) on the interior cells
        Compute norm of residual in the pressure equation ||r^it||
        it := it + 1
    Compute u^(n+1) and v^(n+1) in the fluid domain according to (3.34),(3.35)
    Set the values for u^(n+1) and v^(n+1) at the fixed boundary
    Determine the values for u^(n+1) and v^(n+1) along the free boundary of Ω_{t_n}
    Compute the particle positions at time t_{n+1}
    Data output and visualization as needed
    t := t + δt
    n := n + 1
```

Algorithm 2. Extension for free boundary value problems.

Numerical experiments on a radial flow field and the dam problem (see Section 7.1) show convergence rates between $O(\sqrt{\delta x})$ and $O(\delta x)$ for the velocities and the location of the surface. However, the curvature of the surface, which is necessary to simulate surface tension effects, cannot be approximated sufficiently. To cope with such problems, we developed, in [Neunhoeffer, 1997], [Griebel et al., 1997], an extension of the described algorithm which is based on the surface marker method [Chen et al., 1991]. It shows $O(\delta x^2)$ convergence for the velocities and $O(\delta x)$ convergence for curvature; for details see [Neunhoeffer, 1997].

6.4 Implementation

1. The flag array in Section 3.4.2 can be extended by, e.g., adding five new bit positions: one for each cell itself along with one for each neighboring cell. These are set to one if the corresponding cell is empty and to zero in the case of a fluid, surface, or obstacle cell. This results in the following sequence of bits:

1=empty, 0=fluid/surface/obstacle					1=fluid/empty/surface, 0=obstacle				
center	east	west	south	north	center	east	west	south	north

2. Moreover, the following functions, which can be collected in a file `surface.c`, must be added:

   i. `struct particleline *INIT_PARTICLES(int *N,int imax,int jmax, REAL delx,REAL dely,int ppc,char *problem)`: The particles are distributed uniformly in the problem-dependent fluid domain Ω_0 as described at the beginning of Section 6.1. These can be held in a data structure such as `particleline` introduced in Section 4.2.2. Depending on the given problem, different areas of the initial configuration may be distinguished by inserting the particles into separate lists and later displaying the particles in each list with a different color or symbol during visualization.

      A pointer to the first particle list is returned.

      The parameter `ppc` (*particles per cell*) passes the number of potential initial particle positions per cell. For square cells ($\delta x = \delta y$), `ppc` should be the square of an integer. This function is called from the initialization section of `main`.

   ii. `void MARK_CELLS(int **FLAG,int imax,int jmax,REAL delx, REAL dely,int N,struct particleline *Partlines)`: In a *loop over all particles*, all cells containing at least one particle are assigned the flag `C_F`, whereas the remaining nonobstacle cells receive the flag `C_X`.

      This is followed by a *loop over all cells* which determines the surface cell types and assigns the appropriate flags (see Section 6.1). This function is called at the beginning of the time-stepping loop in `main`.

   iii. `void SET_UVP_SURFACE(REAL **U,REAL **V,REAL **P,int **FLAG,int imax,int jmax,REAL Re,REAL delx,REAL dely,REAL delt)`: The boundary values for u, v, and p are calculated as described in Section 6.2 depending on the flag value of the cells. This requires first determining all velocity values after which the pressure values can be calculated. This function is called in `main` following `MARK_CELLS`.

3. In addition, the following functions must be modified.

   i. In `main`: The integer parameter `ppc` as well as the calls to the functions specified above must be inserted. Moreover, the particles must be advanced at the end of the time-stepping loop by a call to `ADVANCE_PARTICLES`.

   ii. In `COMP_FG`: F and G are not calculated for edges bordering on empty cells.

   iii. In `COMP_RHS`: The right-hand side of the pressure equation is calculated only for interior cells.

   iv. In `POISSON`: In the SOR iteration, the pressure values are updated only in the interior cells. Similarly, only contributions from the interior cells may be summed in the residual calculation. The normalization of the residual is obtained through division by only the number of interior cells.

v. In ADAP_UV: Analogously to F and G, the velocities u and v at the new time level are only calculated at edges not bordering on empty cells or obstacle cells, i.e., at edges whose velocities have not yet been reset by SET_UVP_SURFACE, SETBCOND, or SETSPECBCOND, respectively.

Chapter 7

Example Applications for Free Boundary Value Problems

7.1 The Breaking Dam

A common test problem for programs which simulate free boundary value problems is that of the "breaking dam" [Browne, 1978], [Harlow & Welch, 1965], [Hirt & Nichols, 1981], in which a vertical wall confining a fluid to one side of a basin is suddenly removed, thereby allowing the fluid to spread out. In our example the fluid is initially confined to a rectangular area on the left of the computational domain extending across one-fifth of its width and its entire height. The fluid can flow out of the domain through the right boundary, while a free-slip condition is imposed on the left and on the bottom.

The following parameters were used in the simulation the results of which are shown in Figure 7.1:

```
imax = 50,      jmax = 20,     xlength = 10.0,  ylength = 4.0,
tau = 0.5,      delt = 0.04,   t_end = 5.0,
eps = 0.001,    omg = 1.7,     gamma = 0.5,     itermax = 500,
GX = 0.0,       GY = -1.0,     Re = 10.0,
UI = 0.0,       VI = 0.0,      PI = 0.0,        ppc = 16,
wW = 1,         wE = 3,        wS = 1,          wN = 3.
```

The figures show the particle positions at different times.

7.2 The Splash of a Liquid Drop

Next, we consider a droplet of fluid falling into a fluid-filled basin (cf. [Harlow & Shannon, 1967a], [Harlow & Shannon, 1967b], [Nichols & Hirt, 1971]). The initial configuration can be seen in the first picture ($t = 0$) of Figure 7.2. The basin occupies the lower half of the full domain $\mathcal{G} := [0, a] \times [0, b]$. The radius of the drop is $b/10$, its center is located at the coordinates $(a/2, 2b/3)$, and the drop has an initial velocity of $u_0 = 0, v_0 = -2$. We set the remaining parameters to

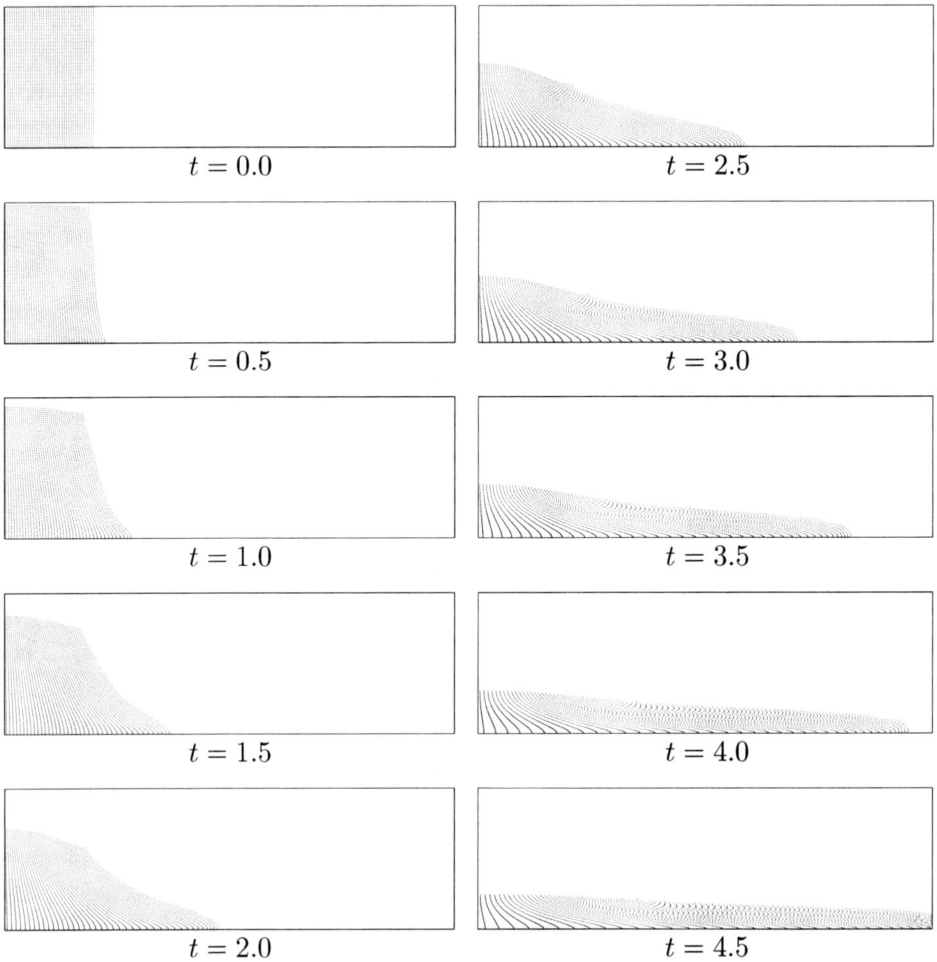

FIG. 7.1. *Breaking dam, time evolution at* $Re = 10$.

```
imax = 40,      jmax = 30,      xlength = 8.0,  ylength = 6.0,
tau = 0.2,      delt = 0.01,    t_end = 10.0,
eps = 0.001,    omg = 1.7,      gamma = 0.5,    itermax = 500,
GX = 0.0,       GY = -1.0,      Re = 40,
UI = 0.0,       VI = 0.0,       PI = 0.0,       ppc = 16,
wW = 1,         wE = 1,         wS = 1,         wN = 3.
```

In Figure 7.2, the particles initially inside the basin are shown as dots, and those comprising the drop as tiny crosses. This distinction of the particles can be maintained for subsequent visualization by keeping the particles in two separate lists of type `Particlelines`.[47]

[47]Because we use cartesian rather than cylindrical coordinates as in [Harlow & Shannon, 1967b], we have actually not simulated the falling of a spherical drop, but that of a "cylindrical" one. Moreover, no symmetries have been exploited.

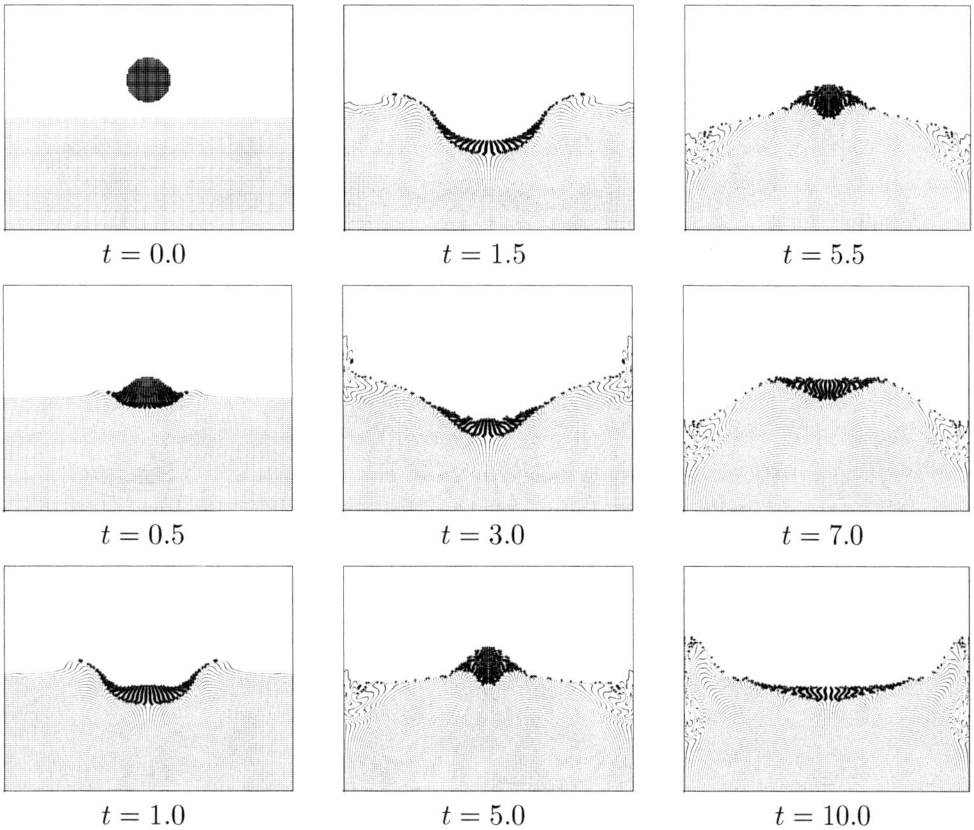

FIG. 7.2. *Splash of a liquid drop, time evolution at Re* = 40.

7.3 Free-Surface Flow over a Step

We can observe a particularly interesting physical phenomenon when considering free-surface flow over a step. We allow the fluid to flow into our initially empty computational domain on the left with constant velocity. The fluid then falls over the edge of the step and drains out on the right, in the course of which an eddy forms between the step and the main flow. After some time, a wave can be observed breaking against the flow direction just beyond the step. This phenomenon, also known as a *hydraulic jump*, can be observed in wild streams or at weirs.[48] We give a brief explanation (cf. [Trefil, 1986]): waves propagate at a certain velocity. For waves propagating on a moving stream in a direction against the flow, there are three different states. If the flow velocity is less than that of the wave (*subcritical flow*), then the wave travels upstream; conversely, if the flow velocity is greater (*supercritical flow*), then the wave is carried downstream. In

[48]These flows, however, are no longer laminar but turbulent.

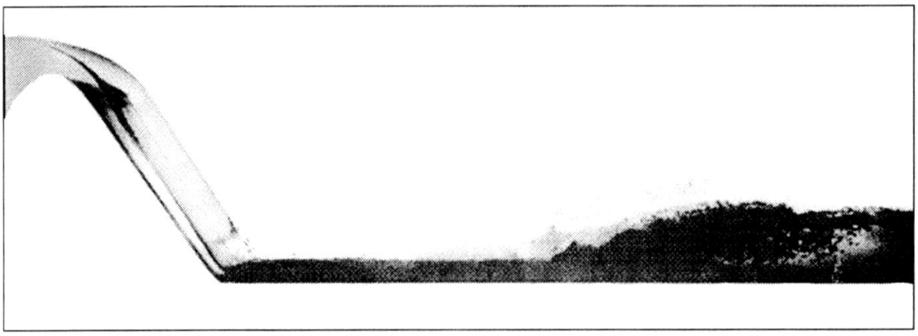

FIG. 7.3. *Backward-breaking wave, physical experiment. (From* [Nakayama, 1988]. *Photograph by K. Matsuo.)*

the case of equal velocities (*critical flow*), the wave stands still. In our case of the flow over a step, the wave speed is nearly constant. The flow velocity, however, is greater just beyond the step than it is further downstream. Therefore, if the flow is subcritical just behind the step and supercritical further on, there must exist a critical point somewhere in-between where the wave remains fixed. This critical point lies precisely at the location where the wave breaks.

A photograph of a physical experiment is shown in Figure 7.3.

The picture sequence in Figure 7.4, which we generated at the suggestion of our colleague Klaus Schlüter, was calculated using the following parameters:

```
imax = 128,    jmax = 32,     xlength = 40.0,  ylength = 10.0,
tau = 0.15,    delt = 0.01,   t_end = 125.0,
eps = 0.001,   omg = 1.7,     gamma = 0.5,     itermax = 500,
GX = 0.0,      GY = -0.5,     Re = 10,
UI = 1.0,      VI = 0.0,      PI = 0.0,        ppc = 16,
wW = 3,        wE = 3,        wS = 2,          wN = 2.
```

The inflow velocity at the left boundary is $u = 1.0$ for $y \in [5, 7.5]$ and the step occupies the region $[0, 5] \times [0, 5]$.

7.4 Injection Molding

An industrial application of free boundary value problems is the injection molding of plastics, which we introduced at the beginning of Chapter 6. For our simulation we will assume that the mold has been evacuated at the outset, thereby allowing us to neglect the outflow of displaced air. The cooling process has been neglected as well. Figure 7.5 shows several snapshots of the injection process.

The parameters used for the computation were

```
imax = 40,     jmax = 30,     xlength = 4.0,  ylength = 3.0,
tau = 0.2,     delt = 0.01,   t_end = 17.8,
eps = 0.001,   omg = 1.7,     gamma = 0.0,    itermax = 100,
GX = 0.0,      GY = 0.0,      Re = 2,
UI = 0.0,      VI = 0.0,      PI = 0.0,       ppc = 64,
wW = 2,        wE = 2,        wS = 2,         wN = 2.
```

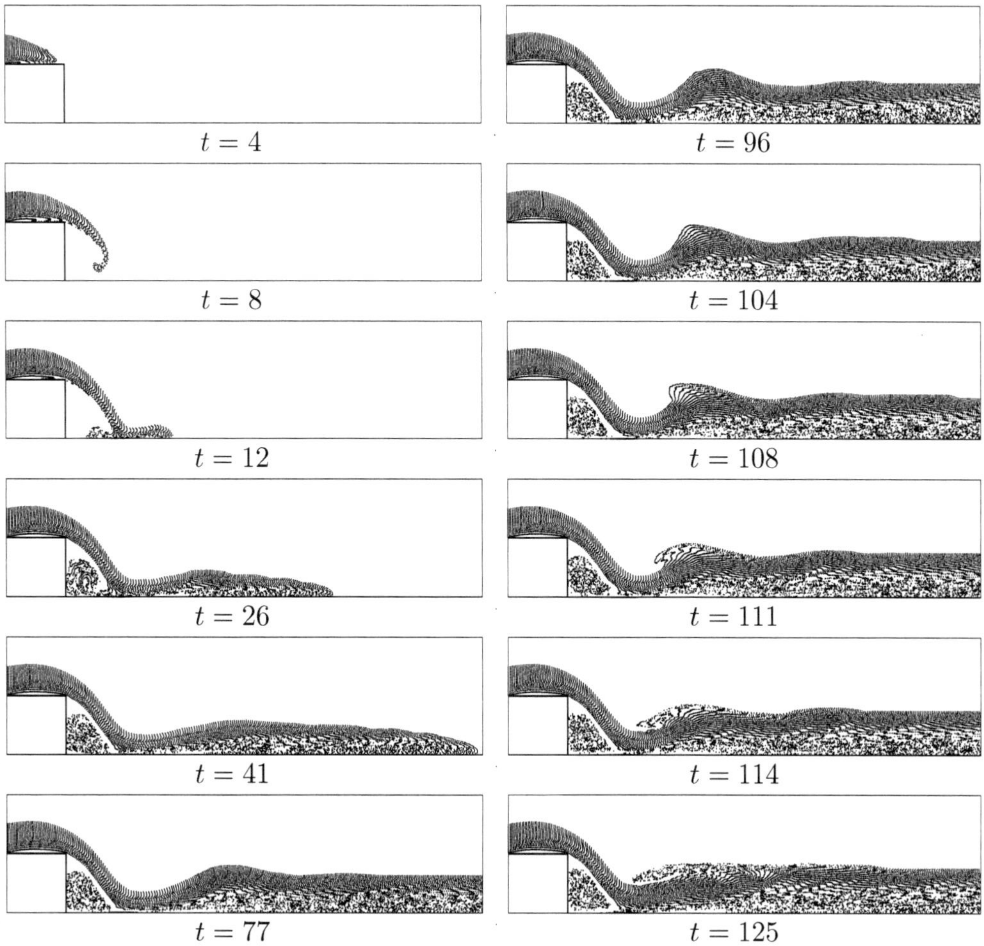

FIG. 7.4. *Backward-breaking wave, time evolution at $Re = 10$.*

The injection velocity at the left boundary between $y = 1.2$ and $y = 1.8$ is $u = 1.0$, $v = 0.0$, and the circular hole has a radius of 0.5 and is centered at $(x, y) = (1.5, 1.5)$.

7.5 Curtain Coating

Coating techniques are another industrial application of free boundary value problems. In this example we introduce a process known as curtain coating, in which three fluid layers are simultaneously applied onto a moving substrate. This technique is used in applications such as the production of photographic film (cf. [Kistler, 1983]). We will restrict ourselves to fluids of the same viscosity and not take into account the hardening process. The initial configuration is depicted in Figure 7.6.

The objective lies in tuning parameters, such as the velocity of the substrate,

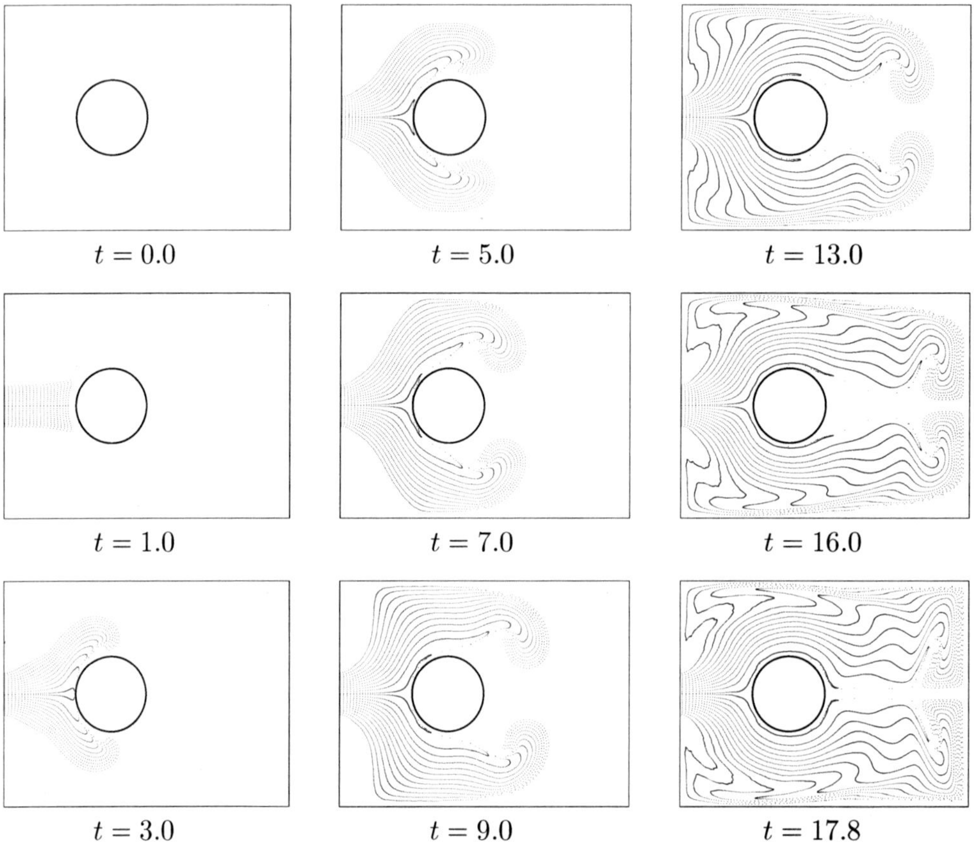

FIG. 7.5. *Injection molding process, Re* $= 2$.

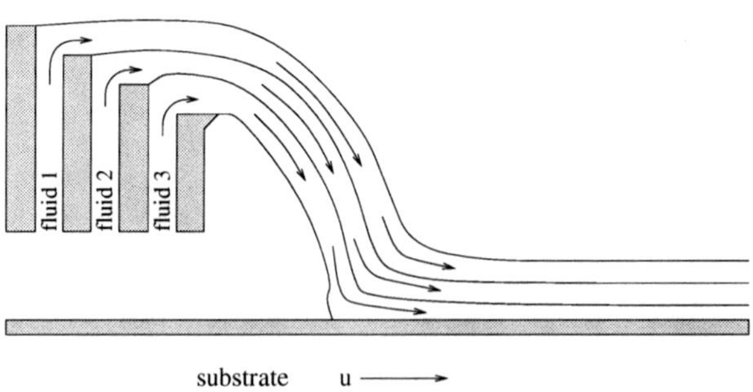

FIG. 7.6. *Curtain coating, schematic diagram.*

the fluids' outflow velocity, or the angle at which the fluids are poured, in order to achieve as smooth and as uniform a coating as possible. The beginning of a coating process is shown in Figure 7.7, in which small waves can still be seen forming on the surface. The substrate velocity here is $u = 2.0$ and the outflow velocity of the fluids is $v = 0.25$. The remaining simulation parameters used are

```
imax = 120,    jmax = 60,    xlength = 24,   ylength = 12,
delt = 0.02,   tau = 0.1,    t_end = 100,
eps = 0.001,   omg = 1.7,    gamma = 0.12,   itermax = 500,
GX = 0.0,      GY = -1.0,    Re = 5,
UI = 1.0,      VI = 0.0,     PI = 0.0,       ppc = 9,
wW = 3,        wE = 3,       wN = 2,         wS = 2.
```

An engineer could now repeat the numerical experiment with little effort using different parameters in order to optimize the coating process.

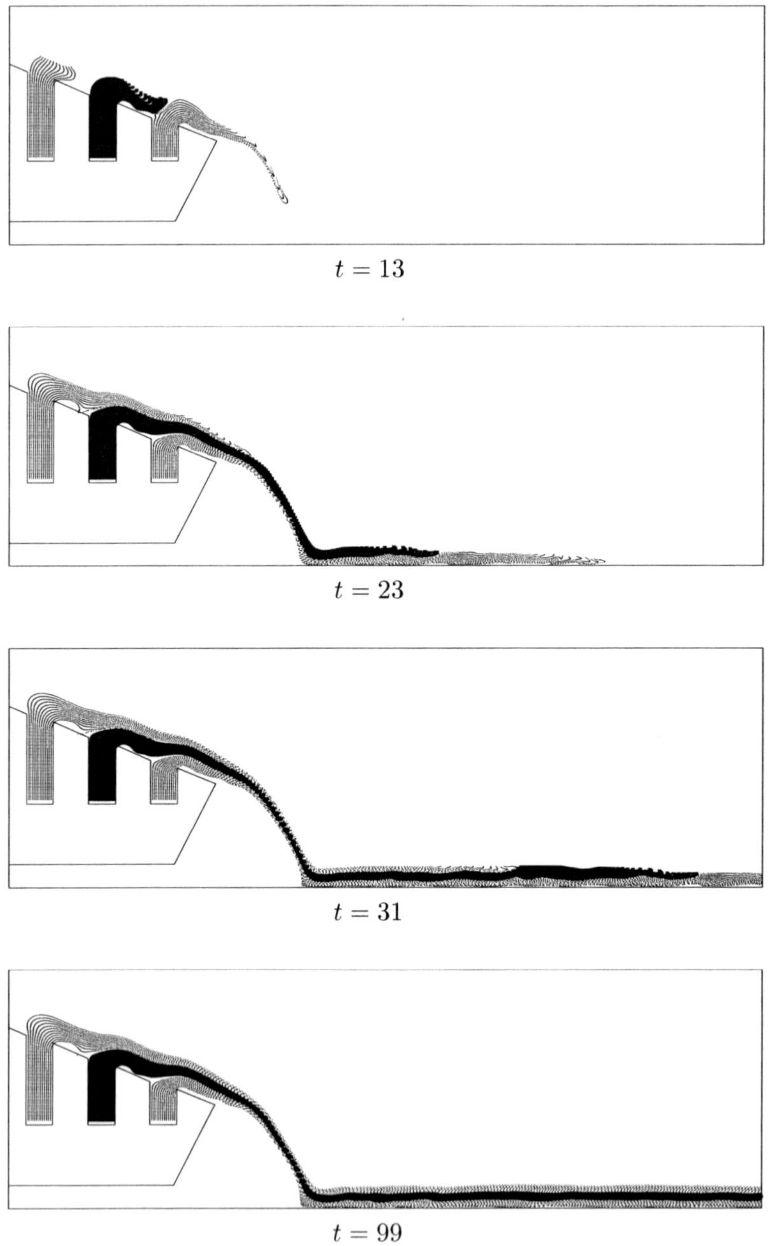

FIG. 7.7. *Beginning of a curtain coating process, time evolution.*

Chapter 8

Parallelization

We now turn our attention to the parallelization of our flow simulation program. Our goal is to reduce the total computing time by dividing the computational work between several processors which, at least to a certain extent, perform their calculations concurrently. In addition, parallel computers are usually equipped with a substantially larger amount of main memory than sequential machines.

One should usually try to make the serial version of an algorithm as efficient as possible before attempting to further accelerate it through parallelization. Such purely algorithmic increases in efficiency could be attained, e.g., by using multigrid methods to solve the pressure equation, by treating the momentum equations (semi)implicitly, or by employing improved discretization methods (higher order methods, adaptive refinement).

Such improvements to our algorithm and their subsequent parallelization, however, are no easy task and lie beyond the scope of this book. This is why, to simplify things, we consider as an example the parallelization of the algorithm as presented in Chapter 3, which uses merely an SOR iteration for solving the pressure equation.

Following a brief overview of the various types of parallel computers and parallel programming environments (cf. [Bräunl, 1993], [Almasi & Gottlieb, 1994]), we describe a parallelization strategy based on the domain decomposition approach. Its implementation is described based on the parallelization software package PVM (for *parallel virtual machine*), available in the public domain, which supplies subroutines for exchanging data between the individual processors forming a parallel computer as well as between individual workstations.

8.1 Parallel Computers and Programming Environments

Vector computers (such as the Cray Y-MP, the Fujitsu S600, or the NEC SX-3R) achieve their high performance by executing similar arithmetic operations on data stored in long vectors in a pipelined fashion. But recent times have seen the success of "true" *parallel computers* consisting of a few dozen to over a thousand powerful processors (usually of reduced instruction set computer (RISC) design) which can compute concurrently and, in addition, must communicate among one another.

Computers belonging to what is known as the MIMD category are characterized by a **m***ultiple* **i***nstruction stream* as well as a **m***ultiple* **d***ata stream*; i.e., the instructions to be executed are distributed among several "streams," the different processors.[49] The same is done with the data to be processed, which is also distributed across several "streams," i.e., across the processors. With regard to physical memory, two types of MIMD computers have established themselves on the market: those with a distributed memory (MIMD *distributed address space machines*), also called *message passing machines*, and those with a global memory (MIMD *global address space machines*), known as *shared memory machines*.[50] [51]

Examples of *distributed memory machines* are the Intel iPSC/860, the Paragon XP/S, the nCube-2, transputer-based systems such as the Parsytec GC, the MEIKO CS-2, IBM's SP1 and SP2, and the CM-5 built by Thinking Machines Corp. Also in this category are groups of workstations connected by a network (Ethernet, HIPPI, FDDI, or ATM switches). The common characteristic of these machines is the distributed manner of programming; i.e., the programs to be executed in parallel are loaded as processes or tasks on each processor. Data allocation occurs locally on each processor and the exchange of data between processors is implemented by explicit statements (*send*, *receive*). Besides the exchange of data, these statements additionally provide the synchronization of the processes in a way similar to the producer-consumer model familiar in computer science (cf. [Bauer & Wössner, 1982, pp. 403 ff.]).

In this context we focus our interest on networks of workstations connected via ethernet and using the PVM system to exchange data.[52] Such groups of workstations may also be used as a parallel computing system, and, due to their general availability and their attractive price-performance ratio, they now offer a true alternative to conventional MIMD computers. The computers connected in this way need not all be of the same architecture; rather, completely heterogeneous networks are possible. Nor is it required that all computers belong to one local network (LAN, or *local area network*); their locations may be distributed over a wide area (WAN, or *wide area network*). With the information highways currently under construction, it will become possible to connect computers across the world to form a virtual parallel computer. Projects connecting computers via the internet are already in progress today [Strumpen, 1995].

[49]There are also computers of SIMD type, which stands for **s***ingle* **i***nstruction,* **m***ultiple* **d***ata*. These consist of inexpensive and relatively weak processors with a rather small instruction set arranged in an array-like manner and tightly synchronized. The resulting large computing power stems from the large number of processors. An example is one of MasPar's computers consisting of up to 65536 processors.

[50]In the latter case, the processors operate in parallel accessing the same common memory. An example of such an architecture is SGI's Power Challenge.

[51]One way of classifying the various types of parallel computers is given by the Erlangen classification scheme; cf. [Bode & Händler, 1983].

[52]Other such packages include Parmacs, Linda, Express, P4, and MPI.

The most powerful parallel computers today, though, are parallel vector computers such as the *Numerical Wind Tunnel* of the National Aerospace Laboratory in Japan, which consists of 140 VPP500 processing units, each with a peak performance of 1.7 GFlops and 256 MBytes of main memory. These combine to form a computer with a theoretical peak performance of 238 GFlops and 36 GBytes of main memory. Besides, new and even more powerful massive parallel machines exist, such as the ASCI Red computer of Sandia National Laboratory, with 7264 Intel processors and a theoretical peak performance of more than 1.4 TFlops. There is also the CP-PACS computer in Tsukuba with 2048 CPUs and 614 GFlops peak performance. For a recent update on the fastest existing supercomputer, see the TOP500 list at http://parallel.rz.uni-mannheim.de.

A more recent approach uses the *virtual shared memory* concept, in which the distribution and exchange of data among the processors is made transparent to the user. The combined memory of all processors is presented to the programmer as a contiguous global address space, i.e., as *one* large memory to which all the computer's processors have simultaneous access. Architectures employing this approach include the KSR-1 and its successor, the KSR-2, from the (now defunct) company Kendall Square Research, the Convex MPP-1, the Cray T3D and T3E, and Silicon Graphics' Origin 2000. Guided by user directives, it is the operating system which performs the automatic distribution of data among the individual processors' local memory. This largely eliminates the need to explicitly program the mapping of processes to processors, local memory allocation, and the interprocess communication steps providing the data exchange. This ease of use in programming, however, is gained at the price of a certain loss of transparency and efficiency in the course of the computation; efficiency depends heavily on the employed caching mechanisms, which are still under development. Hand-tuned code is generally faster.

Up to now, a major problem in the area of parallel computing has been the large number of programming environments and languages, of which there are nearly as many as there are models of parallel computers. While sequential programs may usually be compiled on almost any machine, porting a parallel program from one parallel architecture to another requires substantial effort. The ever shortening development cycles for new parallel computers as well as the strong fluctuation among parallel computer vendors have exacerbated this problem. Moreover, many parallel programs are optimized to match the specific architecture of a given parallel computer.

It is, therefore, absolutely necessary to establish standards, particularly in the area of message passing. This is the goal in the development of the message passing interface (MPI) [Gropp et al., 1994], which benefits from the experience gained from working with PVM, Parmacs, and Express.

8.2 Domain Decomposition as a Parallelization Strategy

An obvious approach to parallelizing numerical algorithms for solving partial differential equations is to divide the underlying domain Ω into subdomains $\Omega_1 \ldots \Omega_N$ and have each process treat one subdomain. The first *domain decomposition method* of this kind was developed by Herrmann Amandus Schwarz [Schwarz, 1890]. For this reason, domain decomposition methods are sometimes known as Schwarz methods. Domain decomposition methods can be classified into overlapping and nonoverlapping variants, depending on whether or not the individual subdomains overlap (see Figure 8.1). We further distinguish between multiplicative methods, in which the subdomains are treated successively, and additive methods, in which they are treated simultaneously. An introduction to domain decomposition techniques may be found, e.g., in [Chan & Mathew, 1994], [Quateroni, 1991], [Smith et al., 1995], or [Hackbusch, 1994, Chapter 11].

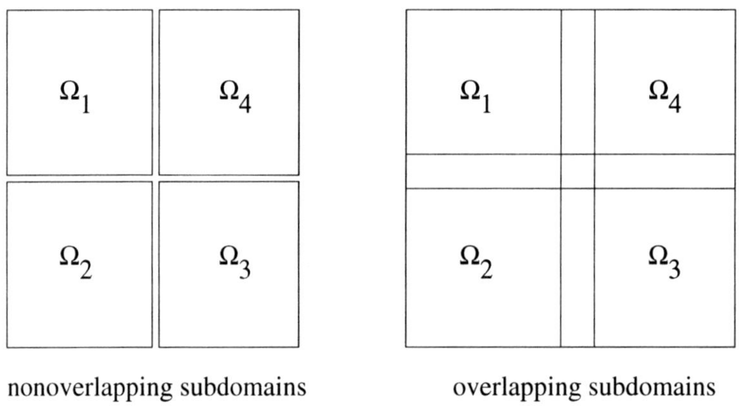

FIG. 8.1. *Domain decomposition methods.*

Each subdomain is thus assigned to a process which computes the unknowns belonging to this subdomain. Each individual process, therefore, no longer requires access to the entire data arrays, and the calculations of an iterative solution algorithm are divided among the processes. If these processes are then assigned to different processors, then the algorithm can be carried out in parallel. Each processor's memory then contains only the data required by those processes running on it. Besides accelerating the computation by parallel execution, parallelization thus results in another advantage: parallel computer systems generally come equipped with considerably more total memory than sequential machines.

To achieve an optimal speedup, the computing load must be distributed across the processors as evenly as possible, assuming that these are all equally powerful. In our application this is accomplished by choosing the subdomains into which the original domain is divided to be of nearly equal size, and hence to contain approximately the same number of unknowns, and by having the same number of processes running on each processor.

To ensure convergence of the resulting algorithm, an exchange of relevant data between processors treating adjacent subdomains (henceforth referred to as *neighboring processes*) becomes necessary at certain times. Generally, it is crucial that only the necessary data be exchanged in as few communication steps as possible, because establishing a connection between processors and exchanging data requires a huge amount of time compared to that required for arithmetic operations on most parallel computers. The increased communication requirements of a parallelized algorithm may somewhat reduce the gain in speed provided by the parallelization of the computations, or it may even completely nullify it.

8.3 Parallelization of the Flow Code

We now turn to the parallelization of the sequential program described in the foregoing chapters.[53] We begin by dividing the base domain for the flow calculation $\mathcal{G}$ along the grid lines into i_{proc} parts in the x-direction and j_{proc} parts in the y-direction, which results in $i_{\text{proc}}\, j_{\text{proc}}$ rectangular subdomains $\Omega_{ip,jp}$, $ip = 1, \ldots, i_{\text{proc}}$, $jp = 1, \ldots, j_{\text{proc}}$. This decomposition is shown in Figure 8.2, in which the grid lines of the staggered grid are shown as dashed lines, and the subdomain boundaries as solid lines. Each process computes the pressure and velocity values in the interior of the associated subdomain. To save an additional communication step, the velocities on each subdomain boundary are computed by both processes involved.

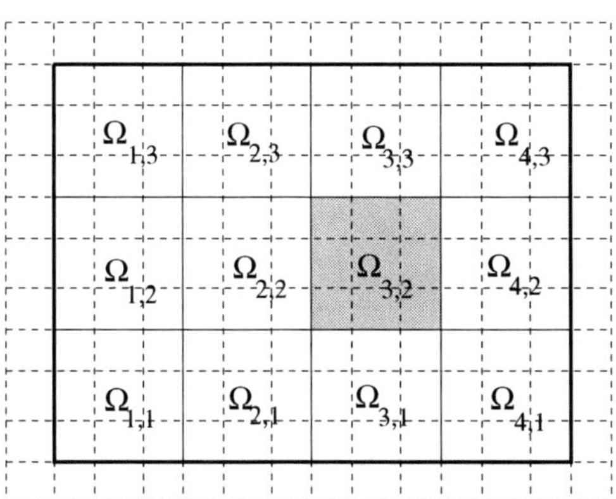

FIG. 8.2. *Subdivision of the domain* Ω ($i_{\text{proc}} = 4, j_{\text{proc}} = 3$).

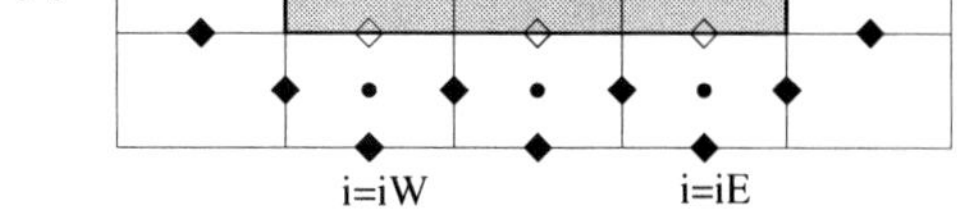

FIG. 8.3. *Unknowns and boundary data for subdomain* $\Omega_{3,2}$.

[53]Further work on the parallelization of flow calculations may be found in, among others, [Pelz et al., 1993], [Simon, 1992], [Fischer & Troger, 1994], and [Schäfer, 1994].

The process treating a subdomain $[(iW-1)\,\delta x, iE\,\delta x]\times[(jS-1)\,\delta y, jN\,\delta y]$[54] will require the values shown in Figure 8.3. Those values marked with light symbols are computed by the process associated with this subdomain, while those marked with dark symbols are only required as "boundary data" for determining the values in the interior and on the boundary of the subdomain. These "boundary values" are—unless they fall outside the original domain Ω—calculated by the processes assigned to the neighboring subdomains. They must therefore be sent by the neighboring processes in a communication phase inside the time-stepping loop.

Following each SOR step in the pressure iteration, the pressure values located in the boundary strips must be exchanged as depicted in Figure 8.4, in order for the computation in the subdomains to proceed with the most current boundary values. The data exchange is performed in four partial steps (to the left, to the right, up, and down). Following the data exchange, the processes are able to compute their respective partial sums of the current residual, which are then sent to one designated process (e.g., the master process) and added there. This process then decides whether to terminate the pressure iteration (tolerance *eps* achieved or maximal number of iterations $it_{\max}$ exceeded) and broadcasts a message to all other processes saying whether to perform the next iteration step or to terminate the pressure iteration.

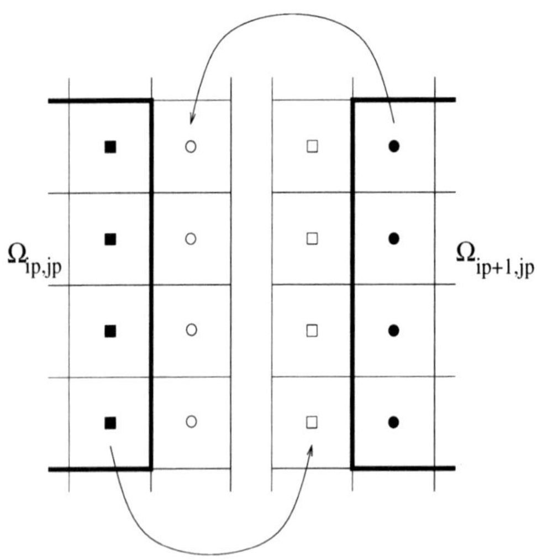

FIG. 8.4. *Exchange of pressure values.*

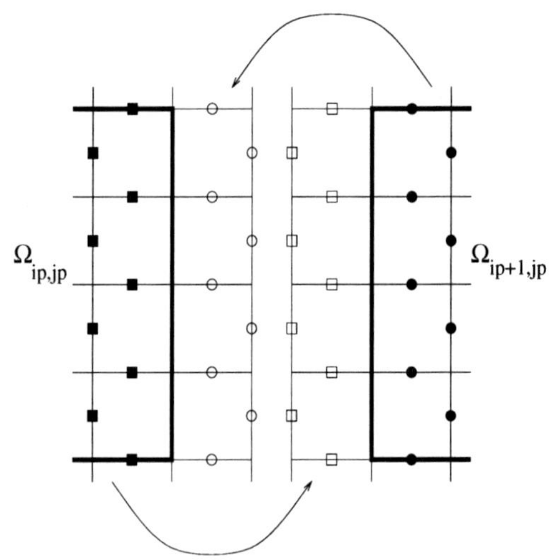

FIG. 8.5. *Exchange of velocity values.*

Since the boundary strips contain the current pressure values at the end of

[54]The quantities iW and iE denote the index of the left (western) and right (eastern) column of cells bounding the subdomain; similarly, jN and jS denote the upper (northern) and lower (southern) bounding rows of cells. For a decomposition into rectangular subdomains, iW, iE, jN, and jS will depend on ip and jp; we suppress this dependence in the interest of clarity.

the pressure iteration, the velocity values on the subdomain boundaries may be updated in ADAP_UV without further communication. Following ADAP_UV, however, the velocities must be exchanged as displayed in Figure 8.5 to enable the calculation of F and G in the next time step.

If a time step size control scheme is used, then—as in the residual calculation—the maximum absolute values of u and v must be determined in each subdomain, whereupon these local maxima are sent to the master process, which determines the global maxima among them, calculates the new δt, and then broadcasts the latter to all processes.

We summarize these considerations in Algorithm 3.

```
Set t := 0, n := 0
Assign initial values to u, v, p
While t < t_end
    Choose δt (according to (3.50) if stepsize control is used)
    Set bounday values for u and v
    Compute F^(n) and G^(n) according to (3.36),(3.37)
    Compute the right-hand side of the pressure equation (3.38)
    Set it := 0
    While it < it_max and ||r^it|| > eps (resp., ||r^it|| > eps ||p^0||)
        Perform an SOR cycle according to (3.44)
        Exchange the pressure values in the boundary strips
        Compute the partial residuals and send these to the master process
        Master process computes the residual norm of the pressure equation
            ||r^it|| and broadcasts it to all processes
        it := it + 1
    Update u^(n+1) and v^(n+1) according to (3.34),(3.35)
    Exchange the velocity values in the boundary strips
    t := t + δt
    n := n + 1
```

Algorithm 3. Parallel version.

When parallelizing the algorithm for solving free boundary value problems, it is necessary to store the positions of all particles currently located in each processor's domain along with the relevant part of the flag array; this includes a boundary strip in addition to the corresponding subarrays of U, V, P, F, G, and RHS. Each process is thus assigned a list of particles. Each time the particles are advanced, it must be determined which particles have crossed which subdomain boundary. To this end, an additional particle list is introduced for each of the four subdomain boundaries. After the particles have been advanced, a particle which has left the subdomain is subsequently deleted from that subdomain's particle list and inserted into the appropriate boundary list. The particle positions in these boundary lists are sent to the process assigned to the subdomain that these particles have entered in the course of the most recent advancing step. Special treatment is necessary for particles moving into a neighboring subdomain which shares only a corner with the subdomain previously occupied (rather than into one of the four subdomains sharing an edge). In this case, a connection must also be established with the corresponding diagonally adjacent subdomain. To avoid additional communication steps, the particles which have traversed a corner may be sent to the left or right neighbor as part of a first communication step exchanging particles across vertical subdomain boundaries; these particles then reach their destination subdomain in a second communication step together with the particles which have crossed horizontal boundaries.

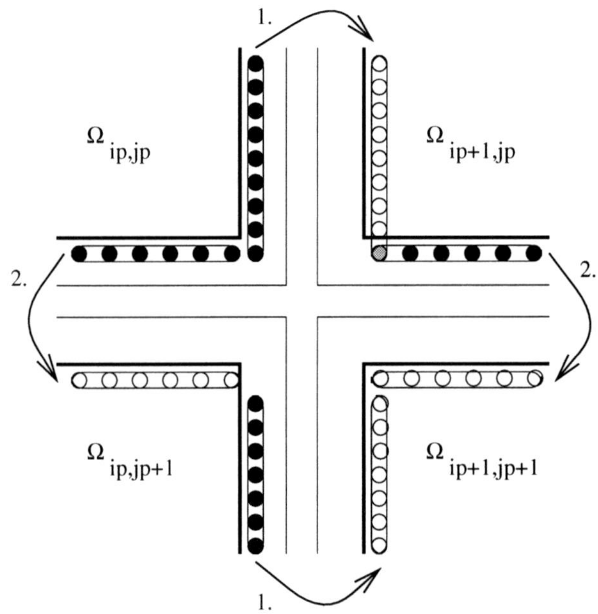

FIG. 8.6. *Exchange of particle positions.*

This technique is shown schematically in Figure 8.6. Here the flow moves from the upper left to the lower right, resulting in particles leaving their subdomains across their right or lower boundaries (black dots). Their positions must now be passed on to the appropriate neighboring processes by sending the particles from left to right in step 1 and from top to bottom in step 2, followed by the insertion of the white points into the particle lists of their new subdomains. A particle having left subdomain $\Omega_{ip,jp}$ across the lower right corner first assumes the intermediate position associated with subdomain $\Omega_{ip+1,jp}$ and marked with a gray dot, before being inserted into the particle list of $\Omega_{ip+1,jp+1}$ following the second communication step.[55] Besides this transport of particles, after each time step, all subdomains must be informed which cells in their boundary strips contain

[55]The particle transport used for visualization via streaklines and pathlines must, of course, be realized as well. When the data is output, all processes must successively write their particle positions out to a file.

particles and which do not. Following the computation of pressure and velocities at the free boundary, these values must once again be exchanged between neighboring subdomains in the manner just described.

Finally, let us note that the parallelization of free boundary value problems for a fixed given decomposition of the original domain into subdomains is not as efficient as the parallelization of flow computations on a fixed given domain, as the load on different processors may vary strongly due to the changing structure of the fluid domain. This requires a dynamic load balancing scheme, in which the subdomains also change with time to maintain an equal distribution of fluid cells among the subdomains, thereby distributing the load on the processors as equally as possible.[56]

8.4 Implementation on a Network of Workstations Using PVM

PVM is a software system allowing a heterogeneous group of computers (workstations, parallel computers, vector computers) to be combined to form a single virtual parallel computer. It thus permits computers already available to be used as one parallel computer. The development of PVM began in 1989 at Oak Ridge National Laboratory supported by the U.S. Department of Energy. Several other institutions are now participating in its ongoing further development. PVM is a public domain product and has found wide distribution in recent years.[57]

PVM consists of two parts: a background process (*daemon*) named `pvmd3`, which combines the individual computers into one virtual parallel computer, and a library, `libpvm3.a`, of interface routines containing functions callable from `C` and `Fortran` for sending and receiving messages, spawning parallel processes, and various other primitives. Details are given in [Geist et al., 1994] and [Geist et al., 1995]. Those functions which we shall use are also described in Appendix A.

The development of the parallel program does not require a different program to be written for each process; rather, a parameter is introduced into a *single* program which serves to identify each process. The domain boundaries as well as the processes assigned to the neighboring subdomains are then inferred from this parameter. At program start, a master process spawns all other processes. This requires the path to the executable program to be known. If all processes can access the same home directory (as is the case in a local network), then the executable program need not be sent to the other processors after compilation.

In the parallelization of our flow code, we follow the approach described in Section 8.2. In the process assigned to subdomain $[(iW-1)\,\delta x, iE\,\delta x] \times [(jS-1)\,\delta y, jN\,\delta y]$, the arrays have the following dimensions:

---

[56]Cf. [Bastian, 1993] for load balancing in adaptive methods.

[57]PVM is available via anonymous `ftp` from `netlib2.cs.utk.edu` in the directory `pvm3`. The current version to date (1996) is PVM 3.3.11.

$$
\begin{array}{ll}
\texttt{P:} & [iW-1, iE+1] \times [jS-1, jN+1], \\
\texttt{U:} & [iW-2, iE+1] \times [jS-1, jN+1], \\
\texttt{V:} & [iW-1, iE+1] \times [jS-2, jN+1], \\
\texttt{F:} & [iW-2, iE+1] \times [jS-1, jN+1], \\
\texttt{G:} & [iW-1, iE+1] \times [jS-2, jN+1], \\
\texttt{RHS:} & [iW, iE] \times [jS, jN], \\
\texttt{FLAG:} & [iW-1, iE+1] \times [jS-1, jN+1].
\end{array}
\tag{8.1}
$$

Implementation

1. First, a file named `hostfile` must be created containing the paths to the daemon as well as to the executable program (see Appendix A).

2. Next, the following functions should be written (and, e.g., collected in a file `parallel.c`):

   i. `void START_PVM(int iproc,int jproc,imax,jmax, int *mytid,int *partid,int *tids,int *iW,int *iE,int *jS,int *jN,int *tW,int *tE,int *tS,int *tN)`: In PVM, each task receives an integer *task identifier* (`tid`) uniquely identifying this process within the PVM system.

   Just as in the PVM example program in Appendix A, the task identifier `mytid` of the process itself, as well as that of the master process `partid`, which will spawn all other processes, is determined.

   The *master process* then spawns the `iproc·jproc-1` remaining processes and sends to these the boundaries `iW,iE,jS,` and `jN` of the subdomain assigned to them as well as the tids `tW,tE,tS,` and `tN` of the processes assigned to the neighboring subdomains. Of course, the master process must also determine these values for itself.

   The *remaining processes* receive their domain boundaries and the tids of their "neighbor processes" from the master process.

   The subdomains should have roughly the same size. This function is called in `main` to start off the initialization phase.

   ii. `void PRESSURE_COMM(REAL **P,int imax,int jmax,int iW,int iE, int jS,int jN,int tW,int tE,int tS,int tN,REAL *bufP)`: The pressure values are exchanged between processes assigned to adjacent subdomains (see Figure 8.4). The following order (or a similar one) should be followed:

send to the left	–	receive from the right,
send to the right	–	receive from the left,
send to the top	–	receive from the bottom,
send to the bottom	–	receive from the top.

Care must be taken that processes with subdomain boundaries falling on those of the base domain $\mathcal{G}$ do not attempt to send or receive data in the corresponding directions. The vector `bufP` can be used whenever data not located in one vector is to be written into the message buffer by a *single* call to `pvm_pk`*type* (see the remark in Appendix A following the description of the PVM routines). This function is called in `POISSON` prior to the calculation of the residual.

iii. `void UV_COMM(REAL **U,REAL **V,int imax,int jmax,int iW,int iE,int jS,int jN,int tW,int tE,int tS,int tN,REAL *bufU,REAL *bufV)`: The velocity values are exchanged between processes assigned to adjacent subdomains (see Figure 8.5). Just as in `PRESSURE_COMM`, the exchange should proceed in four steps; `bufU` and `bufV` have the same function as `bufP`. This function is called at the end of `ADAP_UV`.

iv. `void OUTPUTVEC_PAR(REAL **U,REAL **V,REAL **P,int imax,int jmax,int iW,int iE,int jS,int jN,char *outputfile)`: The data of each process are either written to separate files, which are then suitably combined by the visualization software, or written row by row to one common file, as before.

3. In addition, the following modifications must be introduced into the existing functions from Section 3.3.2:

   i. In `main`, the arrays to be allocated now have the dimensions given in (8.1).

   ii. All loops over i and j may only apply to the local subdomain, i.e., to the local data.

   iii. In `SETBCOND`, `SETSPECBCOND`, and `COMP_FG`, the corresponding boundary values may only be set where the subdomain boundary coincides with the boundary of the base domain or when the subdomain contains an obstacle.

   iv. In `POISSON`, the residual calculation must be modified as explained in Section 8.2.

   v. In `COMP_DELT`, the determination of the next time step length must be modified as described in Section 8.2.

   vi. The newly written functions are to be placed at the locations specified previously.

4. For the parallelization of the solver for free boundary value problems, the following functions must be adapted as well:

   i. In `ADVANCE_PARTICLES`, it must be determined whether particles have left the local subdomain. These are then deleted from the (subdomain's) particle list and inserted into the appropriate boundary list. After the particles have been advanced, the particles in the boundary lists are exchanged again following a fixed given order similar to the one specified in the description of `PRESSURE_COMM`. This must include the separate treatment of particles venturing across corners.

ii. In `MARK_CELLS`, after traversing all particles to divide the cells into fluid cells and empty cells, the values of the flag array in the boundary strips must be sent analogously to the exchange of pressure values.

iii. At the end of `SET_UVP_SURFACE`, the newly determined pressure and velocity values in the boundary strips must be exchanged.

8.5 Measuring Performance

Compared with the serial program, considerable increases in computing speed may be achieved with the parallel program given that the grid is sufficiently fine and enough processors are employed. This increase in performance may be expressed using the terms *speedup* and *efficiency*. These are defined as

- parallel speedup: $S(p) := T(1)/T(p)$,
- parallel efficiency: $E(p) := T(1)/(p \cdot T(p)) = S(p)/p$,

in which p is the number of processors used and $T(p)$ is the execution time of the parallel algorithm running on p processors.[58]

For the driven cavity problem solved on a grid of 1000×1000 cells over 100 time steps, we have obtained the timing results listed in Table 8.1 on a network of 16 HP 9000/720 workstations connected via ethernet on a tree-like network topology using PVM. The measurements were taken immediately before and after the time-stepping loop; the initialization phase involving the setup of the processes and memory allocation is not accounted for.[59]

p	1	2	4	8	16
$T(p)$ in sec.	6130	3375	1861	1175	778
$S(p)$	1.00	1.82	3.29	5.22	7.88
$E(p)$ in %	100.0	90.8	82.3	65.2	49.2

TABLE 8.1. *Parallel execution times, speedup, and efficiency.*

[58] For a fair comparison, $T(1)$ should be the execution time of the fastest *serial* implementation of the algorithm, which contains no parallelization overhead.

[59] In C, the system call `int gettimeofday(struct timeval *tp, struct timezone *tzp)` may be used for timing measurements (the function header is defined in `time.h`). The structure `timeval` consists of the two components `tv_sec`, the number of seconds since January 1, 1970 (the beginning of time for UNIX), and `tv_usec`, containing the number of microseconds passed in addition to these. The local time zone is returned in the structure `timezone`. The function `gettimeofday` must be called once at the beginning and once more at the end of the program segment to be timed. The first timing value is then subtracted from the second.

The time required for communication between the processors of course prevents us from achieving the ideal values of $S(p) = p$ and $E(p) = 100\%$. This deviation from the ideal values actually increases with the number of processors. The goal of parallel program development must therefore be to keep the communication costs and the amount of data to be exchanged as low as possible, particularly when considering that, at the current state of technology, the computing performance of CPUs is increasing considerably faster than the communication performance of available networks.

Chapter 9

Energy Transport

As we have seen in the previous chapters, many types of flows can be treated with the equations we have been using so far, namely, those expressing conservation of mass and momentum. However, one is often interested in also determining the effects of temperature variations on the flow or the transfer of heat within the flow. A thorough understanding of heat transfer, the temperature field, and the associated flow field is of crucial importance in various industrial applications such as energy conversion processes, energy storage, the design of power plants and cooling towers, and crystal growth from the liquid phase, as well as in natural and environmental applications such as meteorology, oceanography, and climatology.

Here again, numerical simulation provides an important tool, since experiments in this area are often costly and effort-intensive and do not permit measurements of sufficient precision and resolution.

In this chapter we describe the necessary extensions of the mathematical model and provide the reader with a derivation of the new equations in the first two sections. These equations are subsequently discretized, and the numerical algorithm is appropriately modified. We further describe a technique for visualizing heat transport. After some implementation suggestions, we include several numerical examples. We conclude with a description of how the dispersal of chemicals in a flow may be computed, allowing for possible chemical reactions among the constituents, for which we also include a simple simulation example.

A thorough treatment of energy transport, its thermodynamic aspects, and its numerical simulation can be found, e.g., in [Patankar, 1980] or [Bejan, 1984].

9.1 Extending the Mathematical Model by the Energy Equation

The Energy Equation

To include the temperature T in our mathematical model we must now turn to the thermodynamic properties of fluids. The principle of *conservation of energy* will then yield the *energy equation*[60] for *constant thermal diffusivity* α, *negligible viscous dissipation*, and a *heat source* q''':

$$\frac{\partial T}{\partial t} + \vec{u} \cdot \mathrm{grad} T = \alpha\, \Delta T + q''' \,. \tag{9.1}$$

The principal effect of temperature on a fluid results from the fact that changes in temperature cause variations in the fluid's density: when heated, a fluid's volume increases, thus making it lighter and causing it to rise. This results in the occurrence of buoyancy forces which depend on the temperature (*thermal buoyancy forces*). Incorporating yet further effects of temperature on the fluid and the flow leads to (nonlinear) equations which are difficult to treat. Hence simplifications are necessary. The most common approach is to use the *Boussinesq approximation*:[61]

- density is constant except in the buoyancy terms;
- all other fluid properties are assumed constant;
- viscous dissipation is negligibly small.

The first assumption means that the continuity equation retains its incompressible form and that, in the momentum equations (2.16) on p. 17, density varies only in the (gravitational) body force term. The remaining assumptions simplify the equations to the effect that the focus of interest is placed on the thermal buoyancy forces.

The momentum equations (2.16) thus take the dimensional form

$$\varrho_\infty \left(\frac{\partial}{\partial t}\vec{u} + (\vec{u} \cdot \mathrm{grad})\vec{u} \right) + \mathrm{grad} p = \mu \Delta \vec{u} + \varrho(T)\vec{g} \,, \tag{9.2}$$

with dynamic viscosity μ.

A further simplification of the Boussinesq approximation assumes a linear relation between ϱ and T,

$$\varrho(T) = \varrho_\infty (1 - \beta\, (T(\vec{x}, t) - T_\infty)) \,, \quad \text{with } \beta := -\frac{1}{\varrho}\frac{\partial \varrho}{\partial T} \,, \tag{9.3}$$

in which ϱ_∞ and T_∞ are reference quantities and β is the *coefficient of thermal expansion*. The momentum equations may thus be written as

$$\varrho_\infty \left(\frac{\partial}{\partial t}\vec{u} + (\vec{u} \cdot \mathrm{grad})\vec{u} \right) + \mathrm{grad} p = \mu \Delta \vec{u} + \varrho_\infty (1 - \beta\, (T(\vec{x}, t) - T_\infty))\vec{g} \,. \tag{9.4}$$

The Boussinesq approximation contains a coupling between the energy equation (9.1) and the momentum equation (9.4), so that, in this model, a flow may be driven solely by a temperature gradient. If a fluid at rest is isothermal ($T \equiv T_\infty$), then these driving forces are zero; heating from a wall, however, produces thermal buoyancy forces on the order of $|\vec{g}|\beta(T_{\mathrm{heating}} - T_\infty)$.

[60] This is a classical *convection-diffusion equation* stating that temperature is *convected* with the flow ($\rightarrow \vec{u} \cdot \mathrm{grad} T$) but also *diffuses* ($\rightarrow \alpha\, \Delta T$) uniformly in all directions.

[61] This approximation is named after Joseph Boussinesq (cf. [Boussinesq, 1903]) despite the fact that it was already used by A. Oberbeck in [Oberbeck, 1879].

For a derivation of the energy equation and a discussion of the validity of the Boussinesq approximation, we refer to Sections 9.2 and 9.3.[62]

Boundary Conditions

For the temperature, there are essentially two different boundary conditions; to impose these, we divide the boundary into two disjoint components: $\Gamma := \Gamma_1 \cup \Gamma_2$.

- *Dirichlet boundary conditions*

$$T|_{\Gamma_1} = T_1.$$

  The temperature is prescribed at the wall. Using this boundary condition, the heating or cooling of the fluid from a wall may be described (e.g., heating of air in a room by a radiator).

- *Neumann boundary conditions*

$$-\kappa \frac{\partial T}{\partial n}\bigg|_{\Gamma_2} = q_w.$$

  Here κ is the fluid's *thermal conductivity* and q_w is the *heat flux* across the wall. This boundary condition describes how much heat is passed on to the wall by the fluid. This is determined by both the material properties of the wall and the temperature difference across the wall. Thus, hot water in a porcelain teapot ($T_{\text{inside}} > T_{\text{outside}}$) will cool down to room temperature quicker than in a thermos bottle ($q_{\text{porcelain}} > q_{\text{thermos}} > 0$). Conversely, cold water ($T_{\text{inside}} < T_{\text{outside}}$) will also warm up faster in the porcelain teapot ($q_{\text{porcelain}} < q_{\text{thermos}} < 0$).

  When $q_w = 0$, this boundary condition is called *adiabatic*, and it then states that no heat is exchanged across this wall (insulated wall).

Hence, along heated or cooled walls as well as along inflow boundaries, temperature is given explicitly, while, along other fixed walls, the heat flux through these is prescribed. At outflow boundaries the adiabatic boundary condition is imposed.

New Quantities

Adding the energy equation to the Navier–Stokes equations introduces a number of new dimensionless quantities expressing important properties of the fluid and the flow. We first mention the

$$\textit{Prandtl number} \qquad Pr := \frac{\nu}{\alpha}, \tag{9.5}$$

[62]When this model is used, the fluid is sometimes called *Boussinesq-incompressible*. A flow driven only by temperature is called *natural convection*.

which describes the relative strength of the diffusion of momentum to that of heat. In contrast with the Reynolds number Re and the Froude number Fr (cf. (2.20) on page 19), the Prandtl number is entirely a property of the fluid (and not of the flow). Typical values range from 3 to 300 for liquids and from 0.7 to 1.0 for gases.

Besides these, the

$$\begin{aligned} &\textit{Grashof number} \quad Gr := \frac{|\vec{g}|\beta(T_2 - T_1)L^3}{\nu^2} \quad \text{and the} \\ &\textit{Rayleigh number} \quad Ra := Pr\, Gr \end{aligned} \tag{9.6}$$

are often used in this context, in which L denotes a characteristic length scale of the flow. The Grashof number Gr expresses the ratio of buoyancy forces to viscous forces; the Rayleigh number Ra is also known as the *modified Grashof number.*

Finally, we mention the

$$\textit{Nusselt number} \qquad Nu := \frac{Q_{\text{convection}}}{Q_{\text{heat diffusion}}}, \tag{9.7a}$$

another important dimensionless parameter expressing the ratio of convective heat transfer $Q_{\text{convection}}$ to diffusive heat transfer $Q_{\text{heat diffusion}}$ in the vicinity of a wall (boundary layer). For the case of heated lateral walls with Dirichlet temperatures T_l and T_r at the left and right walls, these may be expressed as

$$Q_{\text{convection}} = \kappa \int_0^{L_y} \left(-\frac{\partial T}{\partial x} \right)_{x=0} dy\,, \qquad Q_{\text{heat diffusion}} = \kappa\, L_y(T_l - T_r)/L_x\,,$$

where L_y is the height and L_x the length of the fluid container under consideration. Thus, the Nusselt number satisfies

$$Nu = \frac{\displaystyle\int_0^{L_y} \left(-\frac{\partial T}{\partial x} \right)_{x=0} dy}{L_y(T_l - T_r)/L_x}\,. \tag{9.7b}$$

Analogous formulas hold for other configurations.

Dimensionless Formulation

Following the approach of Section 2.3 we now formulate the two equations (9.4) and (9.1) in terms of dimensionless quantities. To this end, we introduce the dimensionless temperature, the dimensionless coefficient of thermal expansion, and the dimensionless heat source in addition to the dimensionless variables already introduced in (2.18) on page 18:

$$T^* := \frac{T - T_\infty}{T_\Delta}\,, \qquad \beta^* := T_\Delta \beta\,, \qquad q'''^* := \frac{L}{T_\Delta u_\infty} q'''\,, \tag{9.8}$$

where T_Δ is a scaling factor.[63] This brings the momentum equation (9.4) into the form

$$\frac{\partial}{\partial t^*}\vec{u}^* + (\vec{u}^* \cdot \text{grad}^*)\vec{u}^* + \text{grad}^* p^* = \frac{1}{Re}\Delta^*\vec{u}^* + (1 - \beta^* T^*)\vec{g}^* \tag{9.9}$$

and the energy equation into the form

$$\frac{\partial T^*}{\partial t^*} + \vec{u}^* \cdot \text{grad}^* T^* = \frac{1}{Re}\frac{1}{Pr}\Delta^* T^* + q'''^* \tag{9.10}$$

containing the Prandtl number from (9.5). Here, $\nu = \mu/\varrho_\infty$ is the kinematic viscosity from (2.17) on page 17.

The Neumann boundary condition becomes

$$\frac{\partial T^*}{\partial n^*} = -\frac{L\,q_w}{\kappa\,T_\Delta}.$$

If the length L_x of the fluid container is chosen as the characteristic length scale L in (2.18) on page 18, then the Nusselt number Nu in (9.7b) may be written simply as

$$Nu = \frac{1}{L_y}\int_0^{L_y}\left(-\frac{\partial T^*}{\partial x^*}\right)_{x^*=0} dy^*\,.$$

For simplicity, we will again omit the stars in the following.

Examples

The simplest example of a purely temperature-driven flow results when a fluid is heated along one lateral wall and cooled along the opposite wall. The heated fluid expands near the first wall and rises. The cooling at the other wall results in the fluid contracting and thus descending. A single cell of rotating fluid is the result.

Figure 9.1 shows a flow produced by heating at the lower wall and cooling at the upper wall. One observes that an entirely different flow pattern from the one with heated lateral walls is established. Here, cell-like cycles of rotating fluid are formed—the so-called *Rayleigh–Bénard cells*.

9.2 Derivation of the Energy Equation

In this section we will derive the energy equation (9.1) from thermodynamic principles (cf., e.g., [Bejan, 1984]). The reader not interested in the physical background may skip this section.

Just as the Navier–Stokes equations were derived from the conservation laws for mass and momentum in Chapter 2, we obtain the energy equation from the First Law of Thermodynamics expressing the conservation of energy:

[63] For flows with heated or cooled walls, T_Δ is usually chosen as the temperature difference $T_\Delta := T_H - T_C$. The reference temperature T_∞ is then taken as either $T_\infty := T_C$ ($\Rightarrow T^* \in [0,1]$) or $T_\infty := (T_H + T_C)/2$ ($\Rightarrow T^* \in [-0.5, +0.5]$).

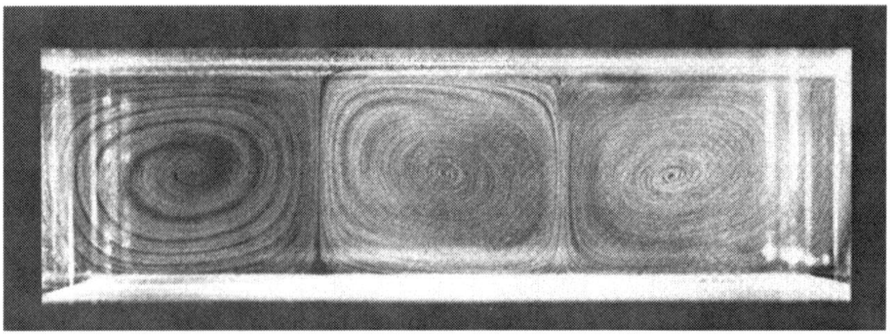

FIG. 9.1. *Flow with heated lower wall, cooled upper wall, and insulated lateral walls (from* [Kessler, 1987]).

The total energy of a system and its surroundings remains constant.

The new quantities now entering the scene are the *internal energy* e, an *internal heat source* $\tilde{q}'''$, and the heat flux vector $\vec{q}''$, which describes the transfer of heat by conduction.

If we take account of all these forms of thermal energy and consider a fixed control volume V, then the energy balance consists of

(a) the rate of change of thermal energy within the control volume V,
(b) internal heat sources (e.g., dissipation of electrical work),
(c) heat transfer across the boundary ∂V of V by the fluid flow,
(d) heat transfer by conduction, and
(e) heat exchanged between V and its surroundings.

These relations, expressed as mathematical formulas, lead to the integral form of the conservation of energy

$$\underbrace{\int_V \frac{\partial}{\partial t}(\varrho e)\, d\vec{x}}_{(a)} = \underbrace{\int_V \tilde{q}'''\, d\vec{x}}_{(b)} - \underbrace{\int_{\partial V} \varrho e \vec{u} \cdot \vec{n}\, ds}_{(c)} - \underbrace{\int_{\partial V} \vec{q}'' \cdot \vec{n}\, ds}_{(d)} + \underbrace{\int_V \boldsymbol{\sigma} \cdot \mathrm{grad}\vec{u}\, d\vec{x}}_{(e)^{64}} . \tag{9.11}$$

Here $\boldsymbol{\sigma}$ is the stress tensor from (2.12): $\boldsymbol{\sigma} = -p\mathbf{I} + \boldsymbol{\tau}$.

To derive the energy equation (9.1), we convert this integral representation into differential form by applying the divergence theorem:

$$\frac{\partial}{\partial t}(\varrho e) = \tilde{q}''' - \mathrm{div}(\varrho e \vec{u}) - \mathrm{div}\vec{q}'' - p\,\mathrm{div}\vec{u} + \boldsymbol{\tau} \cdot \mathrm{grad}\vec{u} . \tag{9.12}$$

[64]This term results from $\int_{\partial V} (\boldsymbol{\sigma}\,\vec{u}) \cdot \vec{n}\, ds = \int_V \mathrm{div}(\boldsymbol{\sigma}\,\vec{u})\, d\vec{x}$ and the splitting $\mathrm{div}(\boldsymbol{\sigma}\,\vec{u}) = \boldsymbol{\sigma} \cdot \mathrm{grad}\vec{u} + \mathrm{div}(\boldsymbol{\sigma}) \cdot \vec{u}$. The second term describes the change in kinetic energy of the fluid contained in V and is neglected. All terms listed here are scalar quantities, in particular $\boldsymbol{\sigma} \cdot \mathrm{grad}\vec{u} = \sum_{i=1}^{3} \vec{\sigma}_i \cdot \mathrm{grad} u_i \in \mathbb{R}$ with $\boldsymbol{\sigma} = (\vec{\sigma}_1, \vec{\sigma}_2, \vec{\sigma}_3)$ and $\vec{u} = (u_1, u_2, u_3)^T$.

After rearranging and introducing the *material derivative*

$$\frac{D}{Dt} := \frac{\partial}{\partial t} + \vec{u} \cdot \mathrm{grad}\,,$$

we obtain the equation

$$\varrho \frac{De}{Dt} + e \underbrace{\left(\frac{D\varrho}{Dt} + \varrho\, \mathrm{div}\vec{u} \right)}_{=0} = \tilde{q}''' - \mathrm{div}\vec{q}'' - p\, \mathrm{div}\vec{u} + \boldsymbol{\tau} \cdot \mathrm{grad}\vec{u}$$

from (9.12).[65] By introducing the enthalpy

$$h = e + p/\varrho\,, \tag{9.13}$$

the material derivative of the internal energy e can be written as[66]

$$\frac{De}{Dt} = \frac{Dh}{Dt} - \frac{1}{\varrho}\frac{Dp}{Dt} + \frac{p}{\varrho^2}\frac{D\varrho}{Dt}\,,$$

and we obtain, after using the continuity equation (2.9) once more along with *Fourier's law*,

$$\vec{q}'' = -\kappa\, \mathrm{grad}T \qquad (\kappa\text{: thermal conductivity}),$$

according to which the heat flux $\vec{q}''$ is proportional to the temperature gradient $\mathrm{grad}T$, the energy equation

$$\varrho \frac{Dh}{Dt} = \tilde{q}''' + \mathrm{div}(\kappa\, \mathrm{grad}T) + \frac{Dp}{Dt} + \boldsymbol{\tau} \cdot \mathrm{grad}\vec{u}\,. \tag{9.14}$$

In order to leave only the temperature T in this equation besides density, pressure, and velocity, we appeal to the thermodynamic relation

$$dh = c_p dT + \frac{1}{\varrho}(1 - \beta T)dp$$

between the rates of change of the fundamental quantities, in which c_p and β are defined as

$$c_p := \left. \frac{\partial h(p,T)}{\partial T} \right|_{p=\mathrm{const}} \quad : \textit{specific heat at constant pressure,}^{67}$$

$$\beta := -\frac{1}{\varrho} \left. \frac{\partial \varrho(p,T)}{\partial T} \right|_{p=\mathrm{const}} \quad : \textit{coefficient of thermal expansion.}$$

[65] The term with the curly underbrace is just the left-hand side of the continuity equation (2.9), which can be seen if one uses the identity $\mathrm{div}(f\vec{g}) = \mathrm{grad}(f) \cdot \vec{g} + f\, \mathrm{div}\vec{g}$ and rearranges some of the terms.

[66] For the material derivative $\frac{D}{Dt}$, the chain rule and product rule both hold as for the ordinary derivative.

The energy equation expressed in terms of the temperature thus reads

$$\varrho c_p \frac{DT}{Dt} = \tilde{q}''' + \mathrm{div}(\kappa\, \mathrm{grad}T) + \beta T \frac{Dp}{Dt} + \boldsymbol{\tau} \cdot \mathrm{grad}\vec{u}\,. \tag{9.15}$$

In our treatment of temperature-related phenomena, we restrict our considerations to fluids of *constant thermal conductivity* κ, *negligible viscous dissipation* $\boldsymbol{\tau} \cdot \mathrm{grad}\vec{u}$, *Boussinesq-incompressible fluids*, and a *heat source* q''', so that the energy equation in the form

$$\frac{\partial T}{\partial t} + \vec{u} \cdot \mathrm{grad}T = \alpha\, \Delta T + q''' \tag{9.16}$$

with the *thermal diffusivity*

$$\alpha = \frac{\kappa}{\varrho_\infty c_p} \tag{9.17}$$

and the scaled heat source

$$q''' = \frac{\tilde{q}'''}{\rho_\infty c_p}$$

finally results.[68]

9.3 On the Validity of the Boussinesq Approximation

As already mentioned in Section 9.1, the density depends on the temperature ($\varrho = \varrho(T)$). However, this also holds for the other thermodynamic quantities, i.e., for the dynamic viscosity $\mu = \mu(T)$, the coefficient of thermal expansion $\beta = \beta(T)$, the specific heat $c_p = c_p(T)$, and hence also for the thermal diffusivity $\alpha = \alpha(T)$. These dependencies are for the most part only empirically quantified, and even these data are incomplete.

In the Boussinesq approximation, it is assumed that the above-mentioned material properties remain constant within certain temperature intervals $T \in [T_C, T_H]$, with the exception of thermal buoyancy forces, which are modeled using a temperature-dependent density $\varrho(T)$. In addition, the density is expanded in a Taylor series around a reference temperature $T_\infty \in [T_C, T_H]$ up to terms of first order (see (9.3)); i.e.,

$$\mu(T) = \mathrm{const}\,, \quad \beta(T) = \mathrm{const}\,, \quad c_p(T) = \mathrm{const}\,, \quad \alpha(T) = \mathrm{const}\,,$$
$$\varrho(T) = \varrho_\infty(1 - \beta(T - T_\infty)) \quad \text{for thermal buoyancy,}$$

[67] The *specific heat* of a fluid is defined as the amount of heat required to raise the temperature of 1 kilogram of this fluid by 1 Kelvin. In this context, the required amounts of heat at constant pressure p (c_p) and at constant volume $1/\varrho$ ($c_v := \partial e(\varrho, T)/\partial T$) turn out to be of particular importance.

[68] Note that $\mathrm{div}(\kappa\, \mathrm{grad}T) = \kappa\, \mathrm{div}(\mathrm{grad}T) = \kappa\, \Delta T$, for $\kappa = \mathrm{const}$.

from which the force density due to thermal buoyancy (as a new source term in the momentum equations) results as $\varrho_\infty \vec{g}\beta(T - T_\infty)$.

These are, however, rather strong simplifications whose validity must be verified. The following criteria are derived in [Gray & Giorgini, 1976]:[69]

$$\left|\frac{\beta|\vec{g}|L}{c_p}\frac{T_\infty}{\Theta}\right| \leq 0.1, \qquad (9.18a)$$

$$\left|\frac{\beta|\vec{g}|L}{c_p}\right| Pr \leq 0.1, \qquad (9.18b)$$

$$|\beta\Theta| = \left|\frac{1}{\varrho}\frac{d\varrho}{dT}\Theta\right| \leq 0.1, \qquad (9.18c)$$

$$\left|\frac{1}{\mu}\frac{d\mu}{dT}\Theta\right| \leq 0.1, \qquad (9.18d)$$

$$\left|\frac{1}{c_p}\frac{dc_p}{dT}\Theta\right| \leq 0.1, \qquad (9.18e)$$

$$\left|\frac{1}{\kappa}\frac{d\kappa}{dT}\Theta\right| \leq 0.1, \qquad (9.18f)$$

$$\left|\frac{1}{\beta}\frac{d\beta}{dT}\Theta\right| \leq 0.1. \qquad (9.18g)$$

Here $\Theta \approx (T_H - T_C)/2$ is the maximal temperature variation around $T_\infty = (T_H + T_C)/2$, and L is a characteristic length of the flow.

As a concrete example we consider water at $T_\infty = 15\,°\text{C}$ $(= 288\text{K})$ and $p_\infty = 1\text{atm}(= 1.01325 \cdot 10^5\text{Pa})$. The required values are taken from [Gray & Giorgini, 1976] and listed in the table below:

$c_p\ [\frac{\text{J}}{\text{kg K}}]$	Pr	$\beta\ [\text{K}^{-1}]$	$\frac{1}{\mu}\frac{d\mu}{dT}\ [\text{K}^{-1}]$	$\frac{1}{c_p}\frac{dc_p}{dT}\ [\text{K}^{-1}]$	$\frac{1}{\kappa}\frac{d\kappa}{dT}\ [\text{K}^{-1}]$	$\frac{1}{\beta}\frac{d\beta}{dT}\ [\text{K}^{-1}]$
4200	8.1	$1.5 \cdot 10^{-4}$	$-2.7 \cdot 10^{-2}$	$-2.4 \cdot 10^{-2}$	$1.7 \cdot 10^{-3}$	$8.0 \cdot 10^{-2}$

For these values, condition (9.18g) restricts the temperature interval the furthest, to

$$\Theta \leq 1.25\,\text{K},$$

and from (9.18b) one obtains the upper bound

$$L \leq 3.5 \cdot 10^4\,\text{m}$$

for the length L.

Finally, condition (9.18a) prescribes a linear relation between the length L and the temperature interval Θ as

$$\frac{L}{\Theta} \leq 9.91 \cdot 10^2\,\frac{\text{m}}{\text{K}}.$$

The resulting region of validity for the Boussinesq approximation is shown in Figure 9.2.

[69]In this paper the dependence of each quantity on the pressure p is considered as well, leading to five extra conditions.

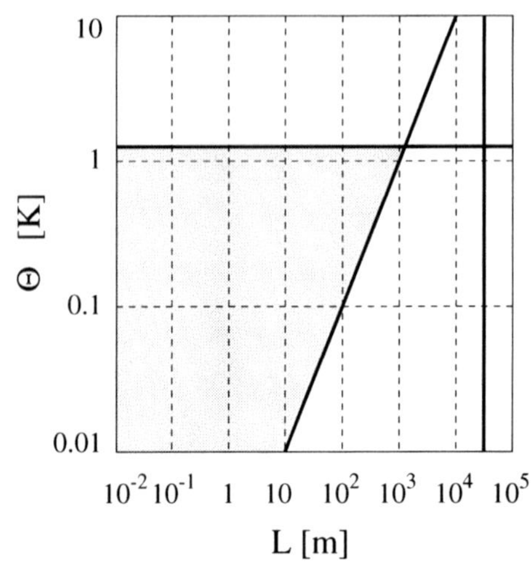

FIG. 9.2. *Region of validity for the Boussinesq approximation for water at* $T_\infty = 15\,°\mathrm{C}$ $(= 288\mathrm{K})$ *and* $p_\infty = 1\mathrm{atm}$.

Since the Boussinesq approximation represents an important tool for analysis and numerical simulation in the area of heat transfer, investigations of this kind give crucial guidelines when treating natural convection.

9.4 Discretization of the Energy Equation and Extension of the Algorithm

To extend our numerical method, the energy equation (9.10) must be discretized on the staggered grid and discrete boundary conditions must be formulated. Furthermore, the stability condition must be adapted and the algorithm appropriately extended.

Discretization of the Energy Equation

As with the momentum equations in (2.1), which were brought to the form (2.2a,b) using the continuity equation, we discretize the dimensionless energy equation (9.10) in the form

$$\frac{\partial T}{\partial t} + \frac{\partial}{\partial x}(uT) + \frac{\partial}{\partial y}(vT) = \frac{1}{Re}\frac{1}{Pr}\left(\frac{\partial^2 T}{\partial x^2} + \frac{\partial^2 T}{\partial y^2}\right) + q''' . \tag{9.19}$$

For the location in the staggered grid where T—and hence the energy equation—is discretized, we choose the cell centers as for the pressure p, and thus the discrete terms of the equation

$$\left[\frac{\partial T}{\partial t}\right]_{i,j}^{(n+1)} + \left[\frac{\partial (uT)}{\partial x}\right]_{i,j} + \left[\frac{\partial (vT)}{\partial y}\right]_{i,j} = \frac{1}{Re}\frac{1}{Pr}\left(\left[\frac{\partial^2 T}{\partial x^2}\right]_{i,j} + \left[\frac{\partial^2 T}{\partial y^2}\right]_{i,j}\right) + q'''_{i,j} \tag{9.20}$$

are given by

$$
\begin{aligned}
\left[\frac{\partial T}{\partial t}\right]_{i,j}^{(n+1)} &= \frac{1}{\delta t}\left(T_{i,j}^{(n+1)} - T_{i,j}^{(n)}\right), \\
\left[\frac{\partial (uT)}{\partial x}\right]_{i,j} &= \frac{1}{\delta x}\left(u_{i,j}\frac{T_{i,j}+T_{i+1,j}}{2} - u_{i-1,j}\frac{T_{i-1,j}+T_{i,j}}{2}\right) \\
&+ \frac{\gamma}{\delta x}\left(|u_{i,j}|\frac{T_{i,j}-T_{i+1,j}}{2} - |u_{i-1,j}|\frac{T_{i-1,j}-T_{i,j}}{2}\right), \\
\left[\frac{\partial (vT)}{\partial y}\right]_{i,j} &= \frac{1}{\delta y}\left(v_{i,j}\frac{T_{i,j}+T_{i,j+1}}{2} - v_{i,j-1}\frac{T_{i,j-1}+T_{i,j}}{2}\right) \\
&+ \frac{\gamma}{\delta y}\left(|v_{i,j}|\frac{T_{i,j}-T_{i,j+1}}{2} - |v_{i,j-1}|\frac{T_{i,j-1}-T_{i,j}}{2}\right), \\
\left[\frac{\partial^2 T}{\partial x^2}\right]_{i,j} &= \frac{T_{i+1,j}-2T_{i,j}+T_{i-1,j}}{(\delta x)^2}, \\
\left[\frac{\partial^2 T}{\partial y^2}\right]_{i,j} &= \frac{T_{i,j+1}-2T_{i,j}+T_{i,j-1}}{(\delta y)^2}.
\end{aligned}
\tag{9.21}
$$

Except for the temperature in the time derivative, all values are those at time t_n, allowing $T^{(n+1)}$ to be explicitly calculated from $T^{(n)}$, $u^{(n)}$, and $v^{(n)}$.

As was done in the discretization of the momentum equations in Section 3.1.2, we have used the *donor-cell scheme* to discretize the convection terms (the second and third terms in (9.21)). Just as in equations (3.19a,b) on page 29, the parameter $\gamma \in [0,1]$ determines a weighted average of discretizing with central differences and the donor-cell discretization. Since the energy equation is discretized at cell centers, the averaging of u and v necessary in the discretization of the momentum equations is not necessary here.

Boundary Conditions

The discrete formulation of the boundary conditions is given here for a rectangular region $\Omega = [0,a] \times [0,b]$ only. The adaptation to general geometries using the flag array, as in Section 3.4, can be done analogously to the boundary conditions already treated in Section 3.4. In edge cells (those bordering on only one fluid cell) the following formulas are used, and in corner cells (those bordering on two fluid cells) one obtains two conditions which can be averaged in the cell.

1. *Dirichlet boundary condition:* The continuous boundary condition, e.g., at a left boundary $T\big|_{\Gamma_l} = T^{0,y}(y)$, is satisfied in the discrete case by averaging:

$$\frac{T_{0,j}+T_{1,j}}{2} = T^{0,y}((j-0.5)\delta y).$$

The discrete boundary conditions at the different boundaries thus read

$$
\begin{aligned}
T_{0,j} &:= 2T^{0,y}((j-0.5)\delta y) - T_{1,j} && \text{for } j = 1,\ldots,j_{\max}\,, \\
T_{i_{\max}+1,j} &:= 2T^{a,y}((j-0.5)\delta y) - T_{i_{\max},j} && \text{for } j = 1,\ldots,j_{\max}\,, \\
T_{i,0} &:= 2T^{x,0}((i-0.5)\delta x) - T_{i,1} && \text{for } i = 1,\ldots,i_{\max}\,, \\
T_{i,j_{\max}+1} &:= 2T^{x,b}((i-0.5)\delta x) - T_{i,j_{\max}} && \text{for } i = 1,\ldots,i_{\max}\,.
\end{aligned}
\tag{9.22}
$$

Compare also the formulas (3.22) and (3.23).

2. *Neumann boundary condition:* Again using the left boundary as an example,

$$
\left.\frac{\partial T}{\partial n}\right|_{\Gamma_l} = T_n^{0,y}(y) \quad \longrightarrow \quad \frac{T_{0,j} - T_{1,j}}{\delta x} = T_n^{0,y}((j-0.5)\delta y)\,,
$$

we arrive at the boundary conditions

$$
\begin{aligned}
T_{0,j} &:= T_{1,j} + \delta x \cdot T_n^{0,y}((j-0.5)\delta y) && \text{for } j = 1,\ldots,j_{\max}\,, \\
T_{i_{\max}+1,j} &:= T_{i_{\max},j} + \delta x \cdot T_n^{a,y}((j-0.5)\delta y) && \text{for } j = 1,\ldots,j_{\max}\,, \\
T_{i,0} &:= T_{i,1} + \delta y \cdot T_n^{x,0}((i-0.5)\delta x) && \text{for } i = 1,\ldots,i_{\max}\,, \\
T_{i,j_{\max}+1} &:= T_{i,j_{\max}} + \delta y \cdot T_n^{x,b}((i-0.5)\delta x) && \text{for } i = 1,\ldots,i_{\max}\,.
\end{aligned}
\tag{9.23}
$$

Note here that $\partial/\partial n$ is the derivative in the *exterior* normal direction.

Stability Conditions

The stability conditions (3.49) on page 39 are extended by yet another one:

$$
\frac{2\delta t}{Re\,Pr} < \left(\frac{1}{(\delta x)^2} + \frac{1}{(\delta y)^2}\right)^{-1}, \tag{9.24}
$$

and formula (3.50) for the stepsize control is augmented by this same condition:

$$
\delta t = \tau \,\min\left(\frac{Re\,Pr}{2}\left(\frac{1}{\delta x^2} + \frac{1}{dy^2}\right)^{-1}, \frac{Re}{2}\left(\frac{1}{\delta x^2} + \frac{1}{dy^2}\right)^{-1}, \frac{\delta x}{|u_{\max}|}, \frac{\delta y}{|v_{\max}|}\right), \tag{9.25}
$$

in which, as before, $\tau \in]0,1]$ represents a safety factor.

Adaptation of the Momentum Equations and Extension of the Algorithm

The algorithm described in Section 3.2 (in particular, Steps 1 to 3 on page 33) must also be suitably extended. This requires that the quantities F and G from (3.36) and (3.37) on page 34 be augmented by the Boussinesq term:

$$
\begin{aligned}
\tilde{F}_{i,j}^{(n)} &:= F_{i,j}^{(n)} - \beta\frac{\delta t}{2}(T_{i,j}^{(n+1)} + T_{i+1,j}^{(n+1)})g_x\,, \\
\tilde{G}_{i,j}^{(n)} &:= G_{i,j}^{(n)} - \beta\frac{\delta t}{2}(T_{i,j}^{(n+1)} + T_{i,j+1}^{(n+1)})g_y\,,
\end{aligned}
\tag{9.26}
$$

where the treatment of the boundary remains unaltered. The completely discretized momentum equations (3.34) and (3.35) then become

$$u_{i,j}^{(n+1)} = \tilde{F}_{i,j}^{(n)} - \frac{\delta t}{\delta x}(p_{i+1,j}^{(n+1)} - p_{i,j}^{(n+1)}), \qquad (9.27)$$
$$i = 1, \ldots, i_{\max} - 1, \quad j = 1, \ldots, j_{\max},$$

$$v_{i,j}^{(n+1)} = \tilde{G}_{i,j}^{(n)} - \frac{\delta t}{\delta y}(p_{i,j+1}^{(n+1)} - p_{i,j}^{(n+1)}), \qquad (9.28)$$
$$i = 1, \ldots, i_{\max}, \quad j = 1, \ldots, j_{\max} - 1.$$

The algorithm is summarized in Algorithm 4.

```
Set t := 0, n := 0
Assign initial values to u, v, p, T
While t < t_end
    Choose δt (according to (9.25) if stepsize control is used)
    Set the boundary values for u, v, and T
    Compute T^(n+1) according to (9.20)
    Compute F̃^(n) and G̃^(n) according to (9.26)
    Compute the right-hand side of the pressure equation (3.38)
    Set it := 0
    While it < it_max and ||r^it|| > eps (resp., ||r^it|| > eps||p^0||)
        Perform one SOR cycle according to (3.44)
        Compute the residual norm of the pressure equation ||r^it||
        it := it + 1
    Compute u^(n+1) and v^(n+1) according to (3.34),(3.35) using F̃^(n) and G̃^(n)
    t := t + δt
    n := n + 1
```

Algorithm 4. Inclusion of the temperature equation.

9.5 Implementation

To incorporate temperature phenomena into the code,[70] the following quantities and functions must be added or extended.

- Problem-dependent quantities:

`REAL`	`Pr`	Prandtl number Pr,
`REAL`	`beta`	coefficient of thermal expansion β,
`REAL`	`TI`	initial value of temperature.

- Data array of dimension `[0,imax+1]x[0,jmax+1]`:

`REAL`	`**TEMP`	temperature T.

[70] For simplicity we have not included an internal heat source q''' in the implementation.

1. `void COMP_TEMP(U,V,REAL **TEMP,FLAG,imax,jmax,delt,delx,dely,GX, GY,gamma,Re,Pr,beta)`: Here the temperature $T^{(n+1)}$ is computed according to equations (9.20) and (9.21) and stored in the array `TEMP`. This function is called in the main program after the boundary values have been set. It may be added to the module `uvp.c`.
2. The function `READ_PARAMETER` must be extended to also read the values `TI` (initial value of temperature), `Pr` (Prandtl number), and `beta` (coefficient of thermal expansion).
3. The function `COMP_FG` must be extended in accordance with (9.26). The array `TEMP` and `beta` are additional parameters to be passed.
4. The function `INIT_UVP` must also initialize the temperature.
5. The boundary conditions for the temperature T are set in the functions `SETBCOND` and `SETSPECBCOND`. A convenient approach is to set all walls to be adiabatic in `SETBCOND` and to treat heating and cooling in `SETSPECBCOND`, which also evaluates the flag array. The relevant formulas are (9.22) and (9.23).
6. The output of the function `OUTPUTVEC` must be extended by the array `TEMP`.

The main program on page 40 must also be extended accordingly.

We have described the extension of the algorithm from Chapter 3, which does not model free boundaries. As a further exercise, we leave it to the interested reader to combine these two algorithms.[71]

9.6 Visualization of Heat Flow

Besides the (temperature-driven) velocity field, which is usually visualized using streamlines, we now turn our attention to the graphical representation of the temperature distribution and the flow of heat. Contour lines are used to display the temperature field, and to reveal the flow of heat, we construct a new function H. This so-called *heatfunction* is constructed in a manner analogous to how we obtained the stream function ψ, i.e., essentially from the continuity equation (2.2c) on page 12. For this, we rearrange the two-dimensional, steady ($\partial T/\partial t = 0$), heat source-free energy equation (9.16) to obtain a divergence form similar to the continuity equation (2.2c):

$$\frac{\partial}{\partial x}(uT) + \frac{\partial}{\partial y}(vT) = \frac{\partial}{\partial x}\left(\frac{\kappa}{\varrho_\infty c_p}\frac{\partial T}{\partial x}\right) + \frac{\partial}{\partial y}\left(\frac{\kappa}{\varrho_\infty c_p}\frac{\partial T}{\partial y}\right)$$

$$\Longleftrightarrow \qquad \frac{\partial}{\partial x}\left(\varrho_\infty c_p uT - \kappa\frac{\partial T}{\partial x}\right) + \frac{\partial}{\partial y}\left(\varrho_\infty c_p vT - \kappa\frac{\partial T}{\partial y}\right) = 0\,.$$

[71]In the case of greatly reduced gravity (so-called *microgravitation*), or in very thin fluid layers, a new phenomenon occurs. In these cases surface tension dominates. If, in addition, the temperature is nonconstant, so-called *thermocapillary forces* arise on the free surfaces, which again constitute surface tensions. This is known as *Marangoni convection*. These phenomena are important at zero gravity and for very thin fluid layers (*thin films*).

Just as in definition (4.4) of ψ, we define the heatfunction by

$$
\begin{aligned}
\frac{\partial H}{\partial y} &= \varrho_\infty c_p u T - \kappa \frac{\partial T}{\partial x} \quad \text{(net heat transport in } x\text{-direction)}, \\
-\frac{\partial H}{\partial x} &= \varrho_\infty c_p v T - \kappa \frac{\partial T}{\partial y} \quad \text{(net heat transport in } y\text{-direction)}.
\end{aligned} \tag{9.29}
$$

This definition (as well as its derivation) exhibits the analogy to the stream function ψ. In the same way as the stream function describes the transport of mass by the flow,

$$
\begin{aligned}
\frac{\partial \psi}{\partial y} &= u \qquad \text{(net mass transport in } x\text{-direction)}, \\
-\frac{\partial \psi}{\partial x} &= v \qquad \text{(net mass transport in } y\text{-direction)},
\end{aligned}
$$

the heatfunction describes how the flow transports heat.

The heatfunction H also has the property that its level curves, known as *heatlines*, run parallel to the local heat flux.

A dimensionless form of the heatfunction is derived in the same manner from the dimensionless energy equation (9.10). This results in the defining equations

$$
\frac{\partial H^*}{\partial y^*} = Re\, Pr\, u^* T^* - \frac{\partial T^*}{\partial x^*}, \qquad -\frac{\partial H^*}{\partial x^*} = Re\, Pr\, v^* T^* - \frac{\partial T^*}{\partial y^*}, \tag{9.30}
$$

in which

$$
H^* = \frac{H}{\kappa\, T_\Delta}. \tag{9.31}
$$

The first example of Section 9.7 contains two flows with similar velocity fields. Their heatlines, however, reveal a qualitative difference in the two flows.

The heatfunction was first introduced in [Kimura & Bejan, 1983].

Implementation

The program is extended by adding the following quantities and functions.

- Data array:

  `REAL **HEAT` heatfunction H.

1. `void COMP_HEAT(U,V,TEMP,REAL **HEAT,FLAG,Re,Pr,imax,jmax,delx, dely)`: Here the discrete quantity $H_{i,j}$ defined at the upper right corner of cell (i,j) is computed according to the discretization

$$
H_{i,j} = H_{i,j-1} + \delta y \left(Re\, Pr\, u_{i,j} \frac{T_{i+1,j} + T_{i,j}}{2} - \frac{T_{i+1,j} - T_{i,j}}{\delta x} \right), \quad i = 0, \ldots, i_{\max},\; j = 1, \ldots, j_{\max},
$$

of the first defining equation in (9.30). As for the stream function ψ, we set $H_{i,0} = 0$ for all i.
Note: This formula only makes sense when the integration can be performed from one adiabatic wall to another. Otherwise, one may either discretize the second defining equation in (9.30) or solve a Poisson problem for H as in [Kimura & Bejan, 1983].
As was the case for the stream function values $\psi_{i,j}$, we must take care that nothing is added to $H_{i,j}$ inside obstacle cells: $H_{i,j} = H_{i,j-1}$.

2. The output of the function `OUTPUTVEC` must be extended by the array `HEAT`.

This new function `COMP_HEAT` may be included in the module `visual.c`.

9.7 Example Applications

9.7.1 Natural Convection with Heated Lateral Walls

The formation of a natural convection flow for heated lateral walls and adiabatic top and bottom walls (see Figure 9.3) will serve as our first example of temperature-driven flows. To demonstrate the effect of the Rayleigh number *Ra* on the flow, we consider two flows with different Rayleigh numbers. In each case we also display the heat flow by employing the *heatlines* introduced in Section 9.6.

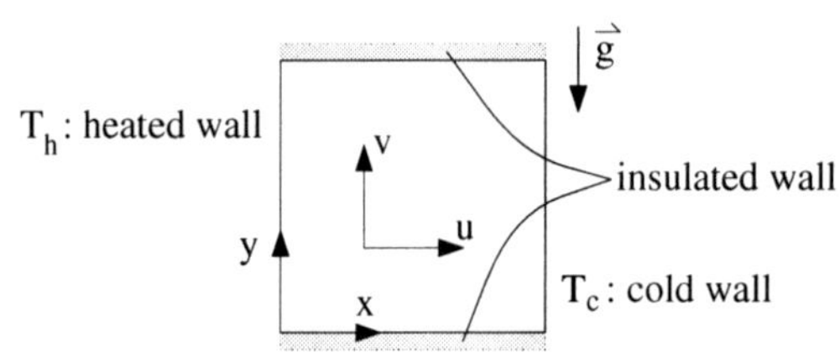

FIG. 9.3. *Flow configuration for heated lateral walls and insulated top and bottom walls.*

1. *Flow with Prandtl number $Pr = 7$ and Rayleigh number $Ra = 1.4 \cdot 10^2$*: Using the program parameters[72]

```
Pr = 7,          Re = 985.7,        beta = 2.1e-4,
GX = 0.0,        GY = -9.706e-2,
xlength = 1,     ylength = 1,
T_H = 1,         T_C = 0
```
(i.e., $T_\infty = T_C$)

and a grid with

```
imax = 50, jmax = 50,
```

[72] An example showing how the required program parameters are computed from the given physical parameters is given in Section 9.7.2.

and imposing no-slip conditions at all four walls, we obtain the flow represented in Figure 9.4. This figure shows the fluid rising along the heated wall and sinking along the cooled wall. The temperature difference is the "engine" driving the flow, in which the processes

$$\begin{array}{ccccc} \text{heating} & \longrightarrow & \text{expansion} & \longrightarrow & \text{rising} \\ \text{cooling} & \longrightarrow & \text{compression} & \longrightarrow & \text{sinking} \end{array}$$

cyclically follow one another.

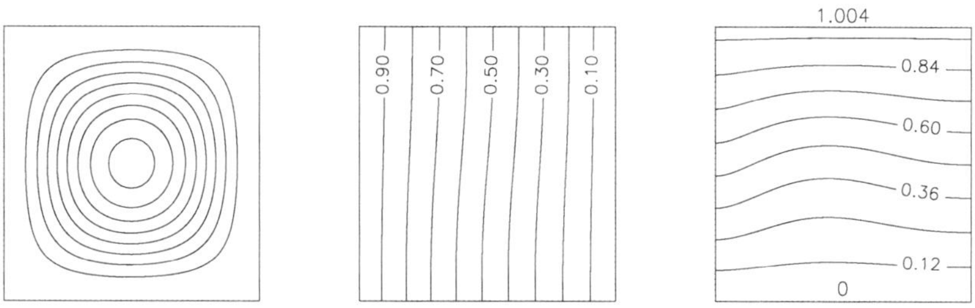

FIG. 9.4. *Streamlines, contour lines of temperature, and heatlines for* $Pr = 7$, $Ra = 1.4 \cdot 10^2$.

The heatlines in Figure 9.4 show heat flowing from the heated wall to the cooled wall.

We also note that, due to the no-slip condition, the value of the heat function along the upper boundary is equal to the Nusselt number of this flow, which is easily verified using the formulas (9.7b) (page 126) and (9.30) (page 137). For this experiment the Nusselt number has the value $Nu = 1.004$.

2. *Flow with Prandtl number* $Pr = 7$ *and Rayleigh number* $Ra = 2.0 \cdot 10^5$: Here the program parameters are

```
Pr = 7,          Re = 11063,     beta = 2.1e-4,
GX = 0.0,        GY = -1.1005,
xlength = 1,  ylength = 1,
T_H = 1,         T_C = 0
```
(also $T_\infty = T_C$),

and the grid is again determined by

```
imax = 50, jmax = 50.
```

The resulting flow is shown in Figure 9.5.

The streamlines indicate that this flow behaves basically the same as that in example 1. The fluid rises near the heated wall and sinks along the cooled wall. But this time internal loops form in the flow as well. The heatlines in Figure 9.5 clearly show that the heat flow proceeds from one wall to the other inside two layers along the boundaries, but also that heat is confined near the center of the domain. This fundamentally distinguishes this flow from the previous one. The Nusselt number for this experiment is $Nu = 5.92$.

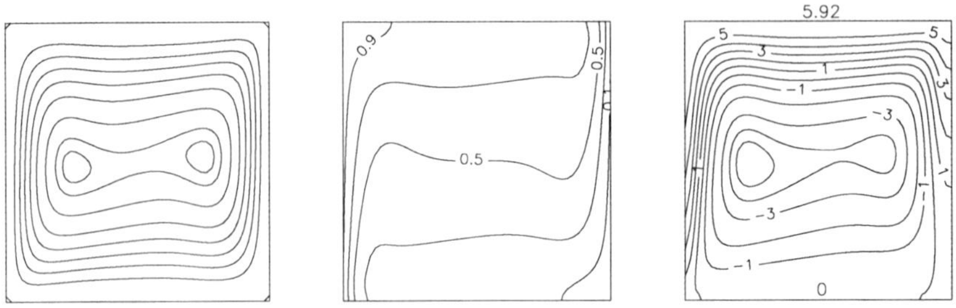

FIG. 9.5. *Streamlines, contour lines of temperature, and heatlines for* $Pr = 7$, $Ra = 2.0 \cdot 10^5$.

Similar examples may be found in [Kimura & Bejan, 1983], in which the authors introduce the heatfunction as a new instrument for analyzing natural convection flows. Similar computations are presented in [Pinelli & Vacca, 1994], in which the flows resulting from heated lateral walls are compared directly with those of heated top and bottom walls.

9.7.2 Buoyancy Flow: Rayleigh–Bénard Convection

In this experiment the two *horizontal walls* are fixed at different temperatures (see Figure 9.6). In contrast with the configuration involving *lateral* walls at different temperatures, for which even very small temperature differences lead to a temperature-driven convection, in this case the temperature difference must exceed a critical value before any flow sets in. A measure which is independent of a particular experiment and the specific fluid under consideration is the Rayleigh number as given in formula (9.6). According to [Bejan, 1984], natural convection will develop only for Rayleigh numbers

$$Ra >\approx 1108.$$

Furthermore, the influence of the lateral walls (carrying no-slip conditions) produces three-dimensional effects, and hence the flow may be approximated as two-dimensional in only two cases: (a) if the lateral walls are far enough apart that their effect may be neglected and (b) if the depth of the horizontal walls is very small (*Hele–Shaw flow*).

Both cases are characterized by the formation of what are called *Rayleigh–Bénard cells*, a cell-like arrangement of eddies of rising and descending fluid (see Figure 9.1). One particular difference between the two- and three-dimensional flows is that mass is exchanged between cells in the latter (cf. the three-dimensional simulation in Section 11.4.4, page 187).

Numerical computations for this case have been performed by, among others, [Kessler, 1987], [Leister, 1994], and [Ozal & Hara, 1995]. Physical experiments may be found in [Wung & Tseng, 1992] and [Jäger, 1982].

We will investigate the formation of Rayleigh–Bénard cells in two simulations.

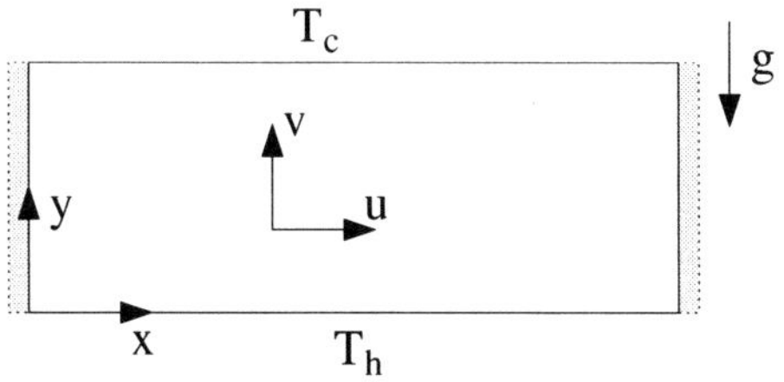

FIG. 9.6. *Flow configuration for heated horizontal walls.*

1. *Flow of glycerine in a container of length × width* $= 38 \times 4$ [cm]: We consider flow in a container 38 cm long and 4 cm wide. The fluid being simulated is to have the properties of glycerine. To find the relevant parameters for this flow, we convert the dimensional physical data of the experiment into the dimensionless quantities used in our code.

   i. *Conversion to dimensionless quantities:* We begin by once more listing all required parameters and normalizations.

$$\text{dimensionless coordinates:} \qquad x^* = \frac{x}{L}, \; y^* = \frac{y}{L}, \; t^* = \frac{u_\infty}{L} t, \tag{9.32a}$$

$$\text{dimensionless states:} \qquad \begin{aligned} u^* &= \frac{u}{u_\infty}, \; v^* = \frac{v}{u_\infty}, \\ p^* &= \frac{p - p_\infty}{\varrho_\infty u_\infty^2}, \; T^* = \frac{T - T_\infty}{T_H - T_C}, \end{aligned} \tag{9.32b}$$

$$\text{dimensionless parameters:} \qquad \begin{aligned} Re &= \frac{\varrho_\infty u_\infty L}{\mu}, \; \frac{1}{Fr^2} = \frac{L\|\vec{g}\|}{u_\infty^2}, \\ Pr &= \frac{\nu}{\alpha}, \; \beta^* = (T_H - T_C)\beta. \end{aligned} \tag{9.32c}$$

The data defining the experiment are

T_C [K]	T_H [K]	T_∞ [K]	$\varrho_\infty \left[\frac{\text{kg}}{\text{m}^3}\right]$	$\mu \left[\frac{\text{kg}}{\text{m}\cdot\text{s}}\right]$	$\beta \left[\frac{1}{\text{K}}\right]$	Pr	$\|\vec{g}\| \left[\frac{\text{m}}{\text{s}^2}\right]$
291.20	294.78	293	1264.02	1.499	$5 \cdot 10^{-4}$	12500	9.81

so that a Rayleigh number of

$$Ra \approx 10^4$$

results (cf. (9.6)).
Normalizing the height to 1, this yields the program parameters

$$\texttt{xlength} = 9.5, \quad \texttt{ylength} = 1,$$

$$\texttt{Re} = 33.73\, u_\infty, \quad \texttt{GY} = -0.3924 \frac{1}{u_\infty^2} \left(\stackrel{\wedge}{=} -\frac{1}{Fr^2} \right),$$

$$\texttt{Pr} = 1.25 \cdot 10^4, \quad \texttt{beta} = 1.79 \cdot 10^{-3}.$$

Since no characteristic velocity is known in advance, we fix it at

$$u_\infty = 1\,\frac{\mathrm{m}}{\mathrm{s}}.$$

This leaves only the ratio between the Reynolds number Re and the Froude number Fr to be chosen.

The experiment is to have a duration of 10^4 sec ($\approx$ 3h), by which (9.32b) leads to a normalized simulation time of

$$\texttt{t\_end} = 2.5 \cdot 10^5.$$

ii. *Simulation results:* The results of the simulation using grids with

`imax = 49, jmax = 5` and `imax = 227, jmax = 21`

are shown in Figures 9.7 and 9.8. These flows resulted in maximal velocities of

$$u_{\max} = u_\infty u^*_{\max} \approx \texttt{4.758e-05}\,\frac{\mathrm{m}}{\mathrm{s}}, \quad v_{\max} = u_\infty v^*_{\max} \approx \texttt{7.781e-05}\,\frac{\mathrm{m}}{\mathrm{s}}$$

and

$$u_{\max} = u_\infty u^*_{\max} \approx \texttt{5.453e-05}\,\frac{\mathrm{m}}{\mathrm{s}}, \quad v_{\max} = u_\infty v^*_{\max} \approx \texttt{8.074e-05}\,\frac{\mathrm{m}}{\mathrm{s}},$$

respectively.

Figures 9.7 and 9.8 clearly show the simulation results for this experiment to depend on the discretization size. In the coarse discretization only 12 Rayleigh–Bénard cells form, compared to 14 in the fine discretization. This indicates that the coarse discretization cannot completely resolve the strucure of the flow. Similar observations are made in [Leister, 1994] as well as [Ozal & Hara, 1995], in which the dependence of the number of cells on the Prandtl and Rayleigh numbers is investigated for small values of the Prandtl number ($Pr \in [0.001, 0.1]$).

2. *Flow of air in a container of length:width* = 4 : 1: Here we consider the flow in an air-filled container with aspect ratio 4:1 (length:width). The physical conditions are determined by fixing the Prandtl number and the Rayleigh number:

$$Pr = 0.72\,, \quad Ra = 30000\,.$$

To determine the parameters required for the simulation, we further require the following quantities:

T_C [K]	T_H [K]	T_∞ [K]	$\varrho_\infty \left[\frac{\mathrm{kg}}{\mathrm{m}^3}\right]$	$\mu \left[\frac{\mathrm{kg}}{\mathrm{m \cdot s}}\right]$	$\beta \left[\frac{1}{\mathrm{K}}\right]$	$\Vert\vec{g}\Vert \left[\frac{\mathrm{m}}{\mathrm{s}^2}\right]$
292.5	293.5	293	1.205	$1.81 \cdot 10^{-4}$	$3.4 \cdot 10^{-3}$	9.8126

FIG. 9.7. *Vector field, streamlines, and temperature contours for the flow of glycerine in a container with aspect ratio* 38:4 (*resolution* 49×5).

FIG. 9.8. *Vector field, streamlines, and temperature contours for the flow of glycerine in a container with aspect ratio* 38:4 (*resolution* 227×21).

Again using the formulas (9.32b,c) and normalizing the height to 1, we obtain the program parameters

$$\texttt{xlength} = 4, \quad \texttt{ylength} = 1,$$
$$\texttt{Re} = 4365\, u_\infty, \quad \texttt{GY} = -0.6432 \frac{1}{u_\infty^2} \quad \left(\hat{=} -\frac{1}{Fr^2} \right),$$
$$\texttt{Pr} = 0.72, \quad \texttt{beta} = 3.4 \cdot 10^{-3}.$$

The simulation results using a grid with

`imax = 65, jmax = 17`

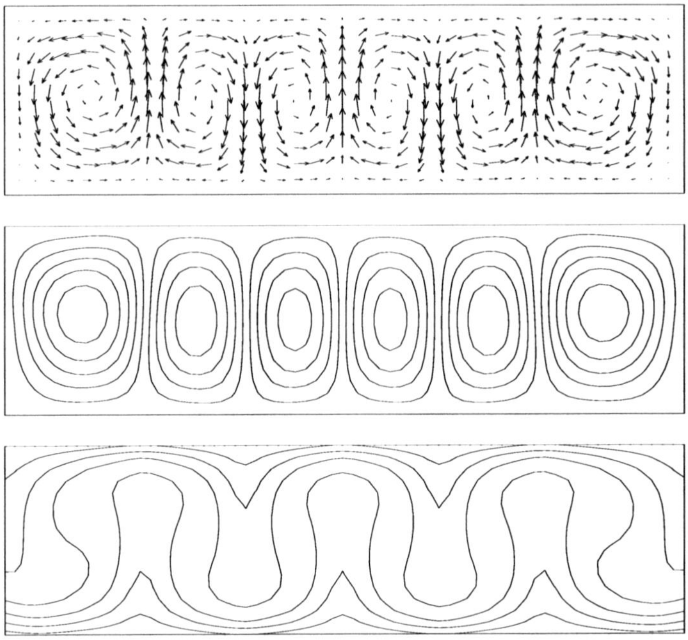

FIG. 9.9. *Vector field, streamlines, and temperature contours for the flow of air in a container with aspect ratio* 4:1.

are shown in Figure 9.9; here the maximal velocities were

$$u_{\text{max}} = u_\infty u^*_{\text{max}} \approx \texttt{0.01304}\,\frac{\text{m}}{\text{s}}, \quad v_{\text{max}} = u_\infty v^*_{\text{max}} \approx \texttt{0.01780}\,\frac{\text{m}}{\text{s}}.$$

This example is treated in a three-dimensional simulation in Section 11.4.4, where the effect of the depth of the fluid container on the number of Rayleigh–Bénard cells is investigated. An extensive discussion of this experiment may be found in [Kessler, 1987].

Various modifications of the configurations described here have been computed. Examples are [Lakhal et al., 1995] and [Khallouf et al., 1995], which study the flow in rectangular containers with periodically heated walls, as well as [Yoo et al., 1994], which gives an extensive treatment of Rayleigh–Bénard cells in concentric cylinders.

9.7.3 Fluid Trap

To approach more realistic configurations such as those found in lakes, in cooling systems of buildings, or in solar panels, we must also consider internal obstacle structures. Here we study three different basic configurations, in each of which the left wall is heated, the right one is cooled, and the remaining walls are insulated.

1. *Vertical internal wall with opening:* When the configuration in Section 9.7.1 (Figure 9.3) is modified by adding a vertical internal wall with an opening, the

fluid can no longer flow unhindered, and heat is exchanged only through this opening. A natural convection flow develops which does not cover the entire domain. Two pools of inactive (*trapped*) fluid form which do not participate in the convective heat transfer. See Figure 9.10.

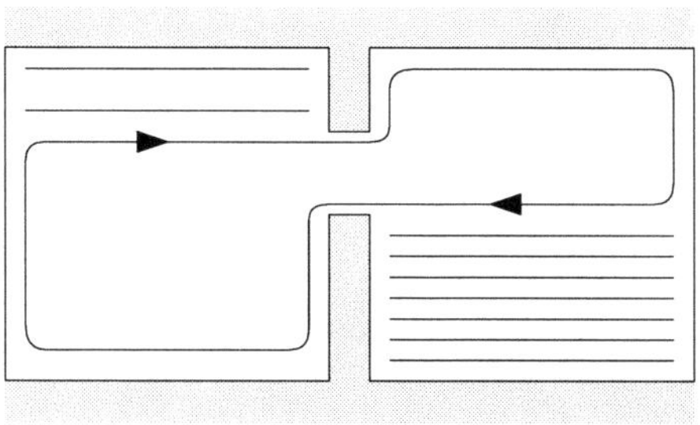

FIG. 9.10. *Configuration with a vertical internal wall.*

The numerical simulation using the program parameters

```
Pr = 7.07,        Re = 9.9621e+4,   beta = 6.3e-4,
GX = 0.0,         GY = -9.81,
xlength = 2,      ylength = 1,
T_H = 0.5,        T_C = -0.5
```
(i.e., $T_\infty = 0.5 \cdot (T_H + T_C)$),

corresponding roughly to a 1 m $\times$ 2 m container filled with 20 °C warm water whose left wall is heated to 21.5 °C and whose right wall is cooled to 18.5 °C, results in the flow picture shown in Figure 9.11.

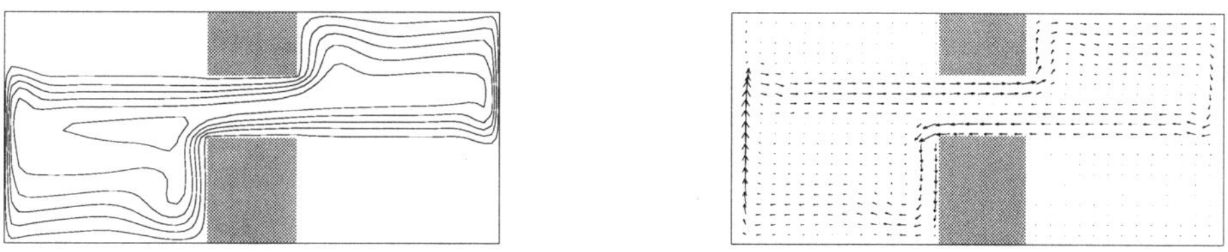

FIG. 9.11. *Streamlines and velocity field of the numerical simulation of the configuration in Figure* 9.10.

One clearly notices the main branch of the flow. In the inactive pools the fluid flows hardly, if at all.

2. *Two vertical internal walls:* When the internal obstacle structure consists of two vertical walls of which one meets only the upper boundary and the other only the lower (and the sum of the heights of the two internal walls exceeds the total container height), it turns out to be essential which side contains the wall adjoining the bottom: if the latter is located on the side of the heated wall (Figure 9.12, left), this results in the formation of a stably stratified region between the internal walls which prevents natural convection

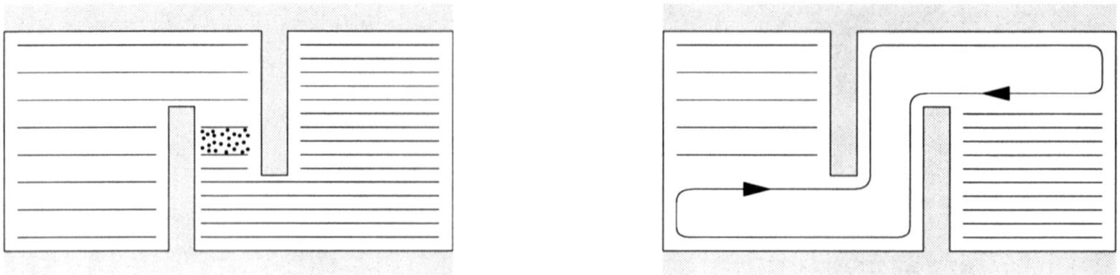

FIG. 9.12. *Configurations with two internal walls.*

from setting in. However, when this wall is on the other side (Figure 9.12, right), then heat is again convected freely.

Numerical results using the same program parameters as in 1 are shown in Figures 9.13 and 9.14.

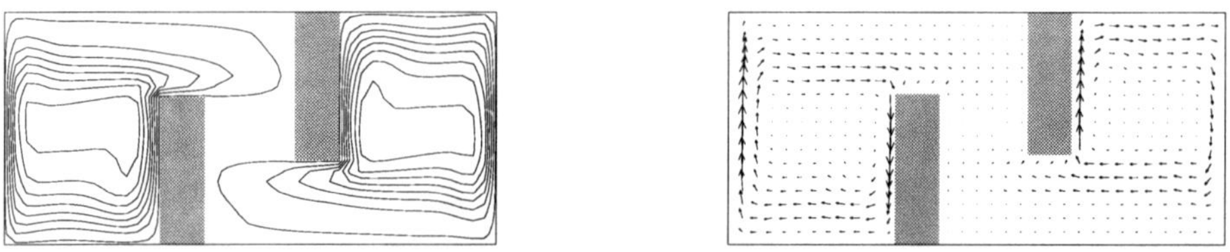

FIG. 9.13. *Streamlines and velocity field of the numerical simulation of the left configuration in Figure* 9.12.

In Figure 9.13 one notices a region between the internal walls across which no convection is taking place. The fluid remains trapped on either side.

FIG. 9.14. *Streamlines and velocity field of the numerical simulation of the right configuration in Figure* 9.12.

Figure 9.14, on the other hand, shows the flow in the nonblocking configuration. Here the fluid can pass between the walls unhindered. The flow in the inactive pools is again negligibly small.

Further investigations on convection flows with internal obstacle structures are carried out in, e.g., [Huang & Vafai, 1994] and [Nag et al., 1994].

9.8 Chemical Transport

9.8.1 Modeling and Discretization

We conclude this chapter with a brief outlook at the basics of treating chemically reacting flows. This topic fits in well at the end of the energy chapter, as the concentration C of a substance dissolved in a fluid satisfies a convection-diffusion equation of the same form,

$$\frac{\partial C}{\partial t} + \vec{u} \cdot \text{grad}C = \lambda \Delta C + Q(t, x, y, \vec{u}, C), \tag{9.33}$$

as the temperature T (cf. equation (9.1)). Here λ is a diffusion coefficient and Q is a source term which may depend on location, time, the velocity, and the concentration itself. In contrast with our treatment of temperature, we will assume for simplicity that the chemical processes under consideration have no effect on the density and hence produce no buoyancy forces. This means that no coupling with the momentum equations is being modeled. The *convective* transport of the substance is described by the term $\vec{u} \cdot \text{grad}C$, and the uniform *diffusive* spreading in all directions, by the term $\lambda \Delta C$.

The discretization of equation (9.33) proceeds along the same lines as that of the energy equation (9.1). As boundary conditions we impose Dirichlet conditions ($C|_{\Gamma_1} = C_0$) along boundary segments Γ_1, along which the substance is being injected, and homogeneous Neumann conditions ($(\partial C/\partial n)|_{\Gamma_2} = 0$) along the remaining portion Γ_2 of the boundary, as in the case of the adiabatic walls in the context of energy transport.

Furthermore, it is easy to extend the model to the case of several mutually dependent chemical species. Here the concentration C_s of each species s is described by its own convection-diffusion equation with a specific source term $Q_s(t, x, y, \vec{u}, \vec{C})$, which may depend on all remaining concentrations. One thus obtains a system of convection-diffusion equations

$$\frac{\partial C_s}{\partial t} + \vec{u} \cdot \text{grad}C_s = \lambda_s \Delta C_s + Q_s(t, x, y, \vec{u}, \vec{C}) \,, \; s = 1, \ldots, s_{\max} \,, \tag{9.34}$$

with

$$\vec{C} = (C_1, \ldots, C_{s_{\max}}) \,.$$

The feedback among the different species is described in the source terms Q_s on the right-hand side. The latter represents the *chemical reactions* occurring in the fluid. The modeling of these chemical reactions by way of the source terms Q_s is usually done using special tables (such as, e.g., *Chemkin* [Kee et al., 1987] or *Jannaf* [Chase, Jr. et al., 1985]), with which the usually nonlinear dependencies of the different species may be calculated in the source terms Q_s.

If these dependencies are strongly nonlinear, as in the case of very rapid chemical reactions, special numerical methods containing specialized discretizations and time step size control (cf. [Oran & Boris, 1987] and [Ern & Giovangigli, 1994]) become necessary.

For the discretization of the boundary conditions, the same statements apply to the modeling of several species as to the modeling of one chemical species.

9.8.2 Implementation

For the implementation of chemical transport we require

- the problem-dependent quantity:

`REAL`	`lambda`	diffusion coefficient λ of species with concentration C,

- the data arrays:

`REAL`	`**C`	concentration C of species,
`REAL`	`**Q`	source term Q of right-hand side of equation (9.33).

Similar to `COMP_TEMP`, a function `COMP_CHEM` is needed and the procedures `READ_PARAMETER` and `INIT_UVP`, as well as the main program, must be appropriately extended.

For the computation of flows with several chemical species the arrays `C` and `Q` may be augmented by a third dimension `[1,...,smax]` for the number of species; the storage for λ is also a one-dimensional array in this case.

We omit these details and leave them to the reader. This algorithm is still, however, far from the treatment of flows involving complex chemical reactions.

9.8.3 Application Example

In this example we qualitatively compute the emission of pollutants in an automobile tunnel. The flow configuration consists of three cars in a channel with no-slip conditions imposed on the upper and lower walls and with inflow and outflow conditions at the walls in the front and rear, respectively (cf. Figure 9.15). This two-dimensional calculation leaves out the cars' wheels in order to allow the fluid to flow under the cars. The cars are standing in the tunnel and exhausts are emitted from the tailpipe of each car. Effects of oncoming traffic and the like are not accounted for.

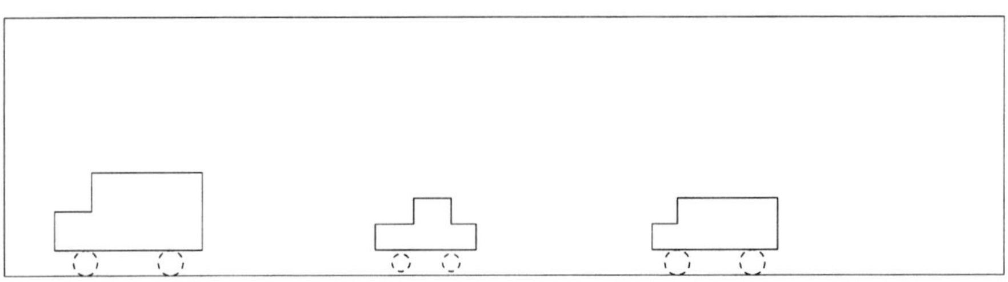

FIG. 9.15. *Flow configuration for pollutant emission in a tunnel.*

Figure 9.16 shows the pollutant spreading diffusively and steadily increasing in concentration in the case of a tunnel ventilated slightly from left to right. The resulting (weak) velocity field is also shown.

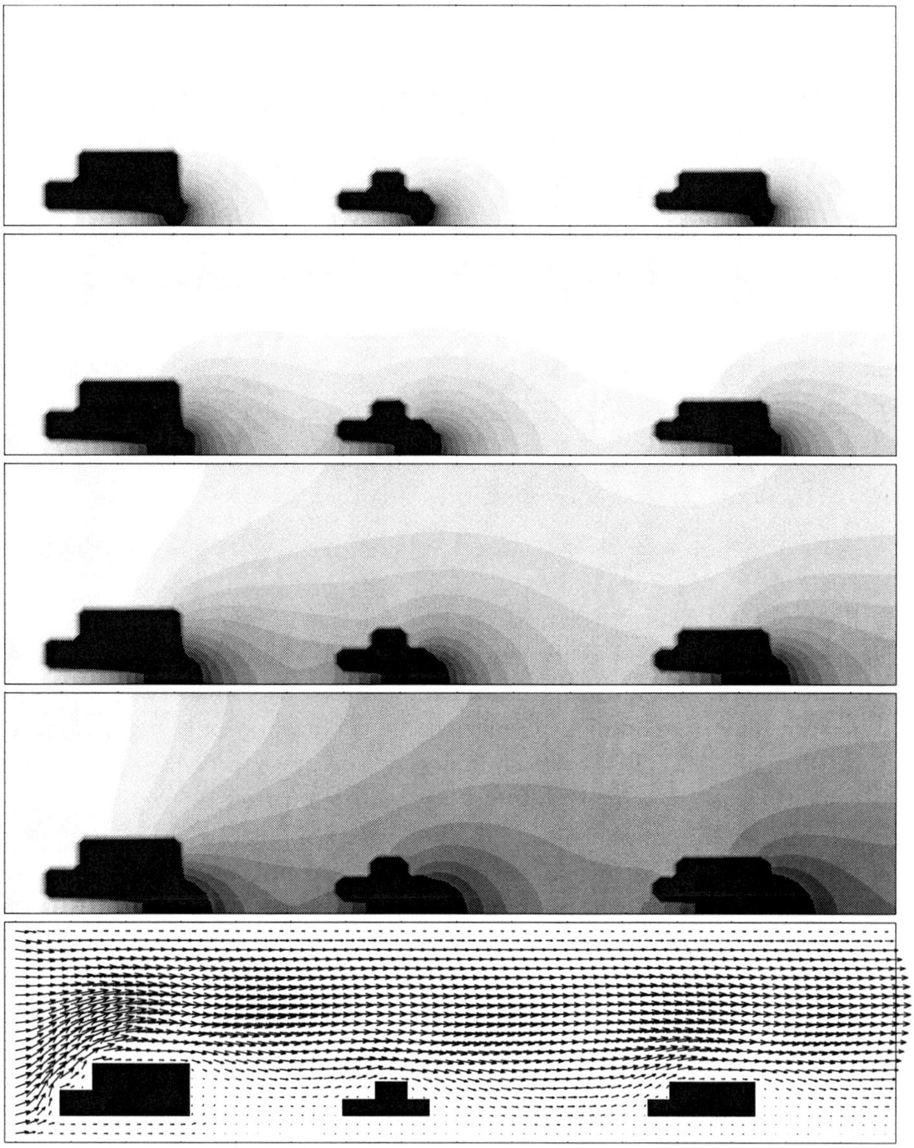

FIG. 9.16. *Slight ventilation at tunnel entrance. Pictures from the top: rise of pollutant concentration with time and associated velocity field.*

In the following, two ventilation options are tested. In one case the slight ventilation of the tunnel is enlarged by a uniform inflow from the tunnel entrance; in the other case three turbines are installed at the tunnel ceiling which accelerate the fluid from left to right. The results are shown in Figures 9.17 and 9.18.

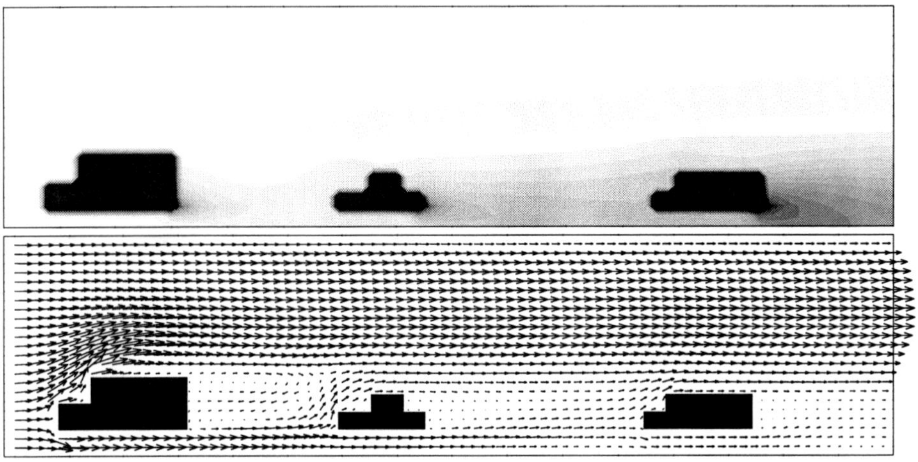

FIG. 9.17. *Tunnel ventilation by inflow at tunnel entrance.*

The first method leads to a noticeable reduction in pollutant concentration (cf. Figure 9.17) and the velocity field contains no eddies which would obstruct a good ventilation. The second method, in contrast, is shown to be less suited. The fluid movement is confined mainly to the ceiling and only very little of the pollutant is carried off. This is accompanied by the development of several eddies in the velocity field, in which pollutants remain captured for quite some time, and which further hinder ventilation of the lower regions of the tunnel.

This example is only of a qualitative nature and not very true-to-life, as no realistic Reynolds number was used and no consideration of turbulence was made. Neither was any account taken of the flow of heat and the resulting buoyancy effects. Nonetheless, our simple simulation indicates that the success of ventilating tunnels with turbines may be questionable, at least for the case of standing traffic.

FIG. 9.18. *Tunnel ventilation by turbines installed at the ceiling.*

Chapter 10

Turbulence

In the previous chapters we have restricted our considerations exclusively to laminar flows. Besides laminar flows, however, it is turbulent flows which most commonly occur in the natural world as well as in technical applications, and it is to their treatment that we now turn our attention.

The analysis and simulation of turbulence ranks among the most difficult tasks in fluid dynamics and can by no means be treated exhaustively within the scope of this book. Nonetheless, we wish to at least address the basics of the mathematical modeling of turbulence as well as the numerical simulation of turbulent flows, along with the difficulties these present.

Following a brief introduction to the physical properties of turbulent flows, we turn to their modeling, for which we employ a widely used turbulence model, the *k-ε model* [Mohammadi & Pironneau, 1993]. Through an averaging process, the *Reynolds equations* are obtained from the Navier–Stokes equations. In addition, two further equations for the *turbulent kinetic energy* k and the *dissipation rate* ε are required. An overview of other turbulence models is followed by the description of the discretization of the k-ε-model, the extension of the algorithm, and the presentation of some numerical results.

10.1 Turbulent Flows

As already mentioned in Section 1.2, flows generally evolve into one of the following types depending on the Reynolds number. We have already investigated some of these in the case of the flow past an obstacle in Section 5.3.

- The flow reaches *steady* state. Either the streamlines closely follow the obstacles due to viscous forces (the viscous forces are stronger than the fluid's inertia) or stable regions of reversed flow develop which remain invariant with time.
- At somewhat higher Reynolds numbers, the regions of reversed flow separate from the obstacle and travel in the main flow direction, while new reversed flow regions form in the obstacle's wake. The flow becomes *periodic.*
- At still higher Reynolds numbers, the eddy formation process as well as the

size of the eddies becomes less and less regular. The flow becomes *quasi-periodic.*

- Finally, for flows at very high Reynolds numbers, an entirely irregular and chaotic flow is established. No periodicity of any kind can be observed any longer; the flow has become *turbulent.*[73]

Turbulent flows possess the following characteristic properties, which distinguish them from laminar flows.

- Turbulent flows are *unsteady* and generally three-dimensional.
- *Irregularity*: Turbulent flows are irregular, chaotic, and unpredictable.
- *Nonlinearity*: Small perturbations of the flow can be amplified to large perturbations which may remain stable for an extended length of time. The reason for this lies in the nonlinearity of the Navier–Stokes equations. The drag is no longer proportional to the velocity, as in the laminar case, but rather to its square, since the flow must supply the kinetic energy for the eddies continually being generated.
- *Diffusivity*: The diffusion rates, i.e., the speeds at which momentum, mass, and heat are spread, are substantially higher in turbulent flows. The reason for this is the transport by means of small and smallest eddies, which spread out isotropically and which are not present in laminar flows.
- *Vorticity*: Turbulent flows are characterized by the occurrence of eddies, whose size may vary over a large range. The largest eddies are on the order of the size of the flow domain; the smallest measure a few millimeters or less. The larger eddies contain the main portion of the flow's energy. This is successively transferred to the smaller eddies, only to be eliminated by viscous dissipation in the smallest eddies.[74]
- *Energy consumption*: Since energy is constantly being eliminated by viscous dissipation (more precisely, converted to heat), maintaining turbulent flows requires a constant supply of energy.[75] This can be achieved, e.g., by persistent oncoming flow at high velocity.

The difference between laminar and turbulent flows may also be thought of as a stability problem: suppose that, in a pipe flow with velocity u, dynamic viscosity μ, and perfectly parallel streamlines, a small perturbation of size r is introduced, which bends the streamlines slightly upward. This reduces the cross-sectional area

[73]Mathematically, turbulence may also be defined in terms of the Fourier spectrum of the flow: if this is continuous, then turbulence has set in. The analysis of the Fourier spectrum permits conclusions on the size of the smallest occurring eddies and hence the number of grid points needed to resolve these.

[74]This process of energy transfer from large eddies to successively smaller eddies is called an *energy cascade*. It is now believed that a small amount of energy is also transferred from the smaller to the larger eddies, a phenomenon known as an *inverse cascade* or a *backscattering effect*.

[75]One then speaks of driven turbulence as opposed to decaying turbulence.

of the flow above the perturbation and, as a result, the velocity correspondingly increases above and decreases below the perturbation. This generates a pressure gradient proportional to ϱu^2, which in turn produces an inertial force accelerating the flow. This force is counteracted by the viscous forces, which are proportional to u/r and act to reduce the velocity gradient in the flow. The relative strength of inertial to viscous forces—this is exactly the Reynolds number $Re = \varrho ur/\mu$ in terms of the size of the perturbation—thus determines whether or not the perturbation will grow, and hence whether the flow remains laminar or becomes turbulent.

In Figure 10.1, we can nicely observe the behavior of a pipe flow at different Reynolds numbers. In the upper picture ($Re = 1500$), the flow is still laminar and dye introduced at the inflow end is carried along with the flow in straight lines. Once a critical Reynolds number is reached (middle picture, $Re = 2340$), the flow starts to become turbulent beyond a certain distance from the inflow entrance. In the lower picture ($Re = 7500$), the flow has become turbulent throughout the entire channel, and the dye is seen to be transported in a very chaotic manner.

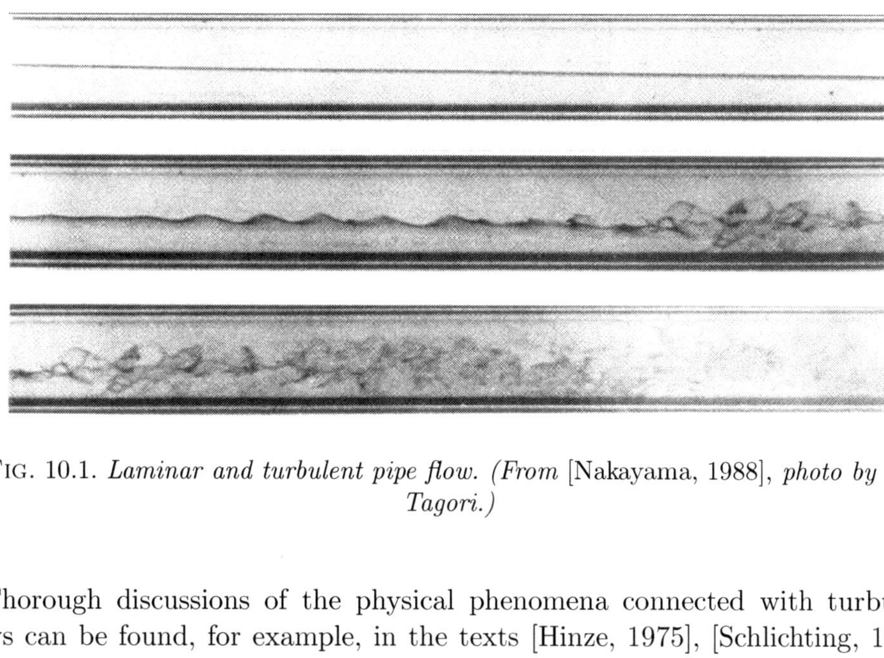

FIG. 10.1. *Laminar and turbulent pipe flow. (From* [Nakayama, 1988], *photo by T. Tagori.)*

Thorough discussions of the physical phenomena connected with turbulent flows can be found, for example, in the texts [Hinze, 1975], [Schlichting, 1982], and [Tennekes & Lumley, 1972].

10.2 Turbulence Modeling

10.2.1 Direct Numerical Simulation and Its Limitations

Since all incompressible flows, whether they possess laminar or turbulent character, are completely described mathematically by the three-dimensional Navier–Stokes equations, the first approach for computing a turbulent flow that springs to mind is the following: discretize and solve the Navier–Stokes equations on a grid sufficiently fine for resolving and representing even the tiniest eddies occurring in the turbulent flow. This is the approach used in what is known as *direct numerical simulation* (DNS).

Kolmogorov's $k^{-5/3}$-law [Kolmogorov, 1942] supposes that the size of the smallest eddies of a turbulent flow depends on the viscosity ν of the fluid and is proportional to $\nu^{3/4}$ (cf., e.g., pp. 23ff in [Mohammadi & Pironneau, 1993]). Since the smallest eddies must be resolved by the grid and the simulation is performed in three space dimensions, this requires $N = O(Re^{9/4})$ grid points in the discretization. In industrial applications such as aerodynamic investigations of automobiles or aircraft, typical viscosity values lie at 10^6 and above. Hence, solving these types of problems properly using direct numerical simulation would require over 10^{13} grid points. Neither existing parallel computers nor computers of the near future can even approximately supply the storage space or the necessary CPU performance demanded by such a simulation.

For this reason, the application of direct numerical simulation is restricted to low Reynolds number problems or problems for which the effects of eddies not resolved by the underlying grid can be neglected.[76] [77]

Figure 10.2 shows an example of turbulent flow through a cylindrical pipe at a maximal Reynolds number of $Re = 6950$ calculated by direct numerical simulation [Eggels et al., 1994]. Analogous results may be obtained using a special memory-conserving method known as the combination technique [Griebel & Huber, 1994].

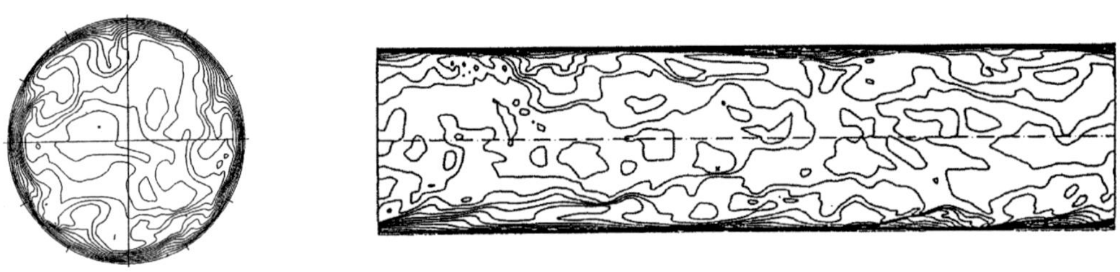

FIG. 10.2. *Turbulent pipe flow, contour lines of velocity in the principal flow direction, cross and longitudinal sections.*

We are thus faced with the following dilemma: on one hand, the computation of turbulent flows at high Reynolds numbers on sufficiently fine grids is not technically feasible now, nor will it be in the near future; on the other hand, using too coarse a grid may incur such large errors in the calculation that useful results

[76]Currently, direct numerical simulation can be applied for Reynolds numbers up to around 10,000.

[77]The highest resolution to date on a cube (1024 × 1024 × 1024 grid points) was achieved by Woodward et al. of the University of Minnesota on 16 SGI computers each containing 20 R4400 processors. These calculations were, however, based on the Euler equations rather than the Navier–Stokes equations.

cannot be obtained.

10.2.2 Basics of Turbulence Modeling

To circumvent this dilemma, the approach to the problem is changed in the following way: in most practical investigations of turbulent flow, the interest is in the macroscopically observed *mean* values rather than in every microscopic detail. Moreover, not every detail of a turbulent flow can even be measured and followed through time. Engineers, rather, are more interested in quantities such as the mean flow rate or the drag of an obstacle. Finally, a certain decomposition into different scales is usually inherent in each flow. One would, as an example, distinguish the main eddies generated by an obstacle from the perturbations resulting from the turbulent oncoming flow, i.e., by the initial and boundary conditions. For this reason, one generally attempts to calculate at least the mean flow using a feasible number of grid points while accounting for the effects of the unresolved eddies by some other suitable means.

This idea, which goes back to Reynolds, lies at the heart of all turbulence modeling. All quantities occurring in the Navier–Stokes equations, i.e., the velocity $\vec{u}$, the pressure p, and the right-hand side $\vec{g}$, are decomposed into a mean part $\vec{U}$, P, and $\vec{G}$, respectively, and the small and smallest variations $\vec{u}\,'$, $p\,'$, and $\vec{g}\,'$ known as *fluctuations*, so that

$$\vec{u} = \vec{U} + \vec{u}\,', \quad p = P + p\,', \quad \vec{g} = \vec{G} + \vec{g}\,'. \tag{10.1}$$

The main part could, for example, be a componentwise statistical, temporal, or spatial average. In the most general formulation, the mean part is formed by applying a filter $\langle\cdot\rangle$, i.e., it holds that

$$\vec{U} := \langle\vec{u}\rangle, \quad P := \langle p\rangle, \quad \vec{G} := \langle\vec{g}\rangle, \tag{10.2}$$

and hence

$$\vec{u}\,' = \vec{u} - \langle\vec{u}\rangle, \quad p\,' = p - \langle p\rangle, \quad \vec{g}\,' = \vec{g} - \langle\vec{g}\rangle. \tag{10.3}$$

The filter is applied to $\vec{u}$ and $\vec{g}$ componentwise.

The filter $\langle\cdot\rangle$ should possess the following properties:[78]

i. $\langle\cdot\rangle$ is a linear operator: for two quantities q and r and a real constant α, there holds $\langle q + r\rangle = \langle q\rangle + \langle r\rangle, \langle\alpha\, q\rangle = \alpha\,\langle q\rangle$.

ii. The time and spatial derivatives commute with the filtering operator: $\langle\partial q/\partial t\rangle = \partial\langle q\rangle/\partial t$, $\langle\partial q/\partial x_i\rangle = \partial\langle q\rangle/\partial x_i$.

iii. Filtered quantities remain invariant under additional filtering: $\langle\langle q\rangle\rangle = \langle q\rangle$.

iv. The filtered product of a quantity with a filtered quantity is equal to the product of the two filtered quantities: $\langle q\langle r\rangle\rangle = \langle q\rangle\langle r\rangle$.

[78] An example of a filter possessing properties i–iv is given by the statistical average $\langle q\rangle_M := \int q(x,t,\omega)d\omega$, in which, e.g., the velocity is regarded as a random variable $u(x,t,\omega)$ with a probability density $d\omega$. The former depends on the random initial condition $u(x,t_0,\omega)$.

We now apply a filtering operator $\langle\cdot\rangle$ to the Navier–Stokes equations. For this we use the vector form of the momentum equations corresponding to (2.2a) and (2.2b).

$$\frac{\partial}{\partial t}\vec{u} + \text{div}(\vec{u}\otimes\vec{u}) + \text{grad}p = \nu\,\Delta\vec{u} + \vec{g}. \tag{10.4}$$

$\vec{a}\otimes\vec{b}$ denotes a tensor with elements $(\vec{a}\otimes\vec{b})_{i,j} := a_i\, b_j$.[79]

Following the application of the filtering operator, the continuity equation becomes

$$0 = \text{div}\langle\vec{u}\rangle = \text{div}\vec{U}, \tag{10.5}$$

and, for the momentum equations, we obtain

$$\frac{\partial}{\partial t}\langle\vec{u}\rangle + \text{div}\langle\vec{u}\otimes\vec{u}\rangle + \text{grad}\langle p\rangle = \nu\Delta\langle\vec{u}\rangle + \langle\vec{g}\rangle. \tag{10.6}$$

Furthermore, it holds that

$$\begin{aligned}\langle\vec{u}\otimes\vec{u}\rangle &= \langle(\vec{U}+\vec{u}\,')\otimes(\vec{U}+\vec{u}\,')\rangle\\ &= \langle\vec{U}\otimes\vec{U}\rangle + \langle\vec{U}\otimes\vec{u}\,'\rangle + \langle\vec{u}\,'\otimes\vec{U}\rangle + \langle\vec{u}\,'\otimes\vec{u}\,'\rangle\\ &= \vec{U}\otimes\vec{U} + \langle\vec{u}\,'\otimes\vec{u}\,'\rangle.\end{aligned}$$

Thus, if a filter possesses properties i–iv, then it can be used to derive from the Navier–Stokes equations[80] what are known as the *Reynolds equations* for the main velocity $\vec{U}$:

$$\frac{\partial}{\partial t}\vec{U} + \text{div}(\vec{U}\otimes\vec{U}) + \text{grad}P - \nu\Delta\vec{U} + \text{div}\langle\vec{u}\,'\otimes\vec{u}\,'\rangle = \vec{G}, \tag{10.7a}$$

$$\text{div}\vec{U} = 0. \tag{10.7b}$$

The averaging of the initial and boundary conditions proceeds analogously. The so-called *Reynolds stresses* are defined as the tensor

$$R(\vec{u}\,') := -\langle\vec{u}\,'\otimes\vec{u}\,'\rangle. \tag{10.8}$$

By filtering the Navier–Stokes equations, we have thus obtained equations which, except for the term $-\text{div}R(\vec{u}\,')$, are identical to the Navier–Stokes equations in the filtered unknowns $\vec{U}$ and P. The term R depends on the velocity fluctuations $\vec{u}\,'$. These are new unknown quantities in addition to $\vec{U}$ and therefore the Reynolds equations no longer constitute a closed system. We now have more unknowns than equations. In order to close the system of equations, we require further equations describing the relations between $R(\vec{u}\,')$ and the quantities $\vec{U}$ and P we seek to determine. The difficulty is that these relations cannot be given in exact form without introducing yet further unknowns. This is known as the *closure problem* in turbulence modeling.

[79] To compute the divergence of a tensor, the divergence operator must be applied separately to each row, resulting in a vector.

[80] If a filter satisfies only properties i and ii, one can still derive terms analogous to the Reynolds stress tensor known as generalized central moments [Germano, 1992], which must then be appropriately modeled.

One is thus forced to appeal to hypotheses and approximations which can generally be justified only heuristically from empirical information and experimental data. The equations thus obtained form the so-called *turbulence model* and serve to close the Reynolds system in an approximate manner.

A fundamental assumption of all turbulence modeling is that the approximation of the Reynolds stress tensor $R(\vec{u}\,') = -\langle \vec{u}\,' \otimes \vec{u}\,' \rangle$ depends *exclusively* on the filtered velocities $\vec{U}$. The former can then be replaced by an expression $\mathcal{R}$ depending on $\vec{U}$ and possibly new quantities, which in turn depend on each other as well as on $\vec{U}$ as specified by additional differential equations.

In a famous hypothesis named in his honor, Osborne Reynolds assumed that

$$R(\vec{u}\,') \approx \mathcal{R}(\text{grad}\vec{U} + \text{grad}\vec{U}^T).$$

This assumption resulted from the observation that turbulence arises mainly in regions with steep velocity gradients. It is, however, not always justified, since, for example, at the leading edge of an aircraft wing the flow is laminar despite the extremely large velocity gradients there.

If one retains this hypothesis anyway, then the functional dependence of the expression $\mathcal{R}$ on $\text{grad}\vec{U} + \text{grad}\vec{U}^T$ may by no means be chosen arbitrarily. It is reasonable, for example, to require that the turbulence model possess the same invariances with respect to the choice of coordinate system as do the Navier–Stokes equations and the Reynolds equations. The latter are invariant under translations and rotations as well as under Galilean transformations.[81]

With these restrictions we obtain, using the abbreviations

$$X := \text{grad}\vec{U} + \text{grad}\vec{U}^T$$

and[82]

$$|X| := \sqrt{\text{trace}(XX^T)},$$

the following expression for $\mathcal{R}$:

$$\mathcal{R}(X) = \eta I + \chi X \quad \text{in two dimensions,} \tag{10.9}$$

$$\mathcal{R}(X) = \eta I + \chi X + \gamma X^2 \quad \text{in three dimensions,} \tag{10.10}$$

in which η, χ, and γ in turn are functions depending only on $|X|$ and $|X^2|$. In [Smagorinsky, 1963],

$$\mathcal{R}(X) = \eta I + ch^2|X|X \quad \text{with } c \simeq 0.01$$

is proposed for the two-dimensional case leading to

[81] Let U denote the mean velocity. The invariance of the Navier–Stokes equations under the transformation $x \mapsto x + Ut, u(x,t) \mapsto u(x + Ut) - U$ is called a Galilean invariance. The new velocity now has mean zero because we are in the frame moving with mean velocity U.

[82] The trace of a square matrix $A := (a_{i,j})$ is defined as the sum of its diagonal elements: $\text{trace}(A) := \sum_i a_{i,i}$.

$$\chi(|X|) := ch^2|X|$$

where $h(x)$ is the local mesh width around the point x of the discretization method being used. The Smagorinsky model is appropriate in the two-dimensional case; it is, however, not sufficient for three dimensions. Moreover, numerical experiments in two dimensions show more or less satisfactory results only for very fine mesh widths h.

10.2.3 The k-ε Turbulence Model

To overcome the disadvantages of the Smagorinsky approach, a number of improved turbulence models have been developed. The most widely used is the k-ε model of [Launder & Spalding, 1972], which was extended to apply to more complex applications in numerous other papers.

The k-ε model is a two-equation model, which means that two additional variables k and ε are introduced to model the Reynolds stresses. Of these, k denotes the *turbulent kinetic energy* and ε the *dissipation rate*:

$$k := \frac{1}{2}\langle|\vec{u}'|\rangle, \qquad \varepsilon := \frac{\nu}{2}\langle|\text{grad}\vec{u}' + \text{grad}\vec{u}'^T|^2\rangle. \tag{10.11}$$

Two supplementary equations for the determination of k and ε are also introduced, thereby closing the system of equations.

The following assumptions underly the k-ε model.

- The Reynolds hypothesis holds, namely, that $R(\vec{u}') = -\langle\vec{u}' \otimes \vec{u}'\rangle$ depends exclusively on k, ε, and grad$\vec{U}$; i.e.,

$$R(\vec{u}') \approx \mathcal{R}(\text{grad}\vec{U} + \text{grad}\vec{U}^T, k, \varepsilon).$$

- The fluctuations $\vec{u}'$ and hence turbulence are isotropic, meaning they are uniform in all directions.
- Convective transport of the fluctuating quantities corresponds to a diffusive transport of the mean quantities.
- $\langle\vec{u}' \otimes \vec{u}'\rangle$ and grad$\vec{U}$ + grad$\vec{U}^T$ are proportional, and the proportionality factor is the *turbulent eddy viscosity*[83]

$$\nu_T := c_\mu \frac{k^2}{\varepsilon}. \tag{10.12}$$

The approximate Reynolds tensor

$$R(\vec{u}') \approx \mathcal{R}(\text{grad}\vec{U}, k, \varepsilon) := -\frac{2}{3}kI + c_\mu \frac{k^2}{\varepsilon}(\text{grad}\vec{U} + \text{grad}\vec{U}^T) \tag{10.13}$$

[83]As a consequence, $\langle\vec{u}' \otimes \vec{u}'\rangle$ and grad$\vec{U}$ + grad$\vec{U}^T$ are parallel in the three-dimensional case. Particularly for the computation of three-dimensional flows, this assumption proves to be overly restrictive in practice.

thus obtained is now substituted in the momentum equations, so that, setting

$$\nu^* := \nu + \nu_T = \nu + c_\mu \frac{k^2}{\varepsilon}, \tag{10.14}$$

we obtain

$$\frac{\partial}{\partial t}\vec{U} + \mathrm{div}(\vec{U} \otimes \vec{U}) + \mathrm{grad}P + \frac{2}{3}\mathrm{grad}k - \mathrm{div}(\nu^*(\mathrm{grad}\vec{U} + \mathrm{grad}\vec{U}^T)) = \vec{G}, \tag{10.15}$$

or, written out in component form in two dimensions,

$$\begin{aligned} \frac{\partial U}{\partial t} + \frac{\partial (U^2)}{\partial x} + \frac{\partial (UV)}{\partial y} + \frac{\partial P}{\partial x} + \frac{2}{3}\frac{\partial k}{\partial x} & \\ - \frac{\partial}{\partial x}\left(2\,\nu^* \frac{\partial U}{\partial x}\right) - \frac{\partial}{\partial y}\left(\nu^* \left(\frac{\partial U}{\partial y} + \frac{\partial V}{\partial x}\right)\right) &= G_x, \end{aligned} \tag{10.16}$$

$$\begin{aligned} \frac{\partial V}{\partial t} + \frac{\partial (UV)}{\partial x} + \frac{\partial (V^2)}{\partial y} + \frac{\partial P}{\partial y} + \frac{2}{3}\frac{\partial k}{\partial y} & \\ - \frac{\partial}{\partial x}\left(\nu^* \left(\frac{\partial V}{\partial x} + \frac{\partial U}{\partial y}\right)\right) - \frac{\partial}{\partial y}\left(2\,\nu^* \frac{\partial V}{\partial y}\right) &= G_y, \end{aligned} \tag{10.17}$$

respectively.

The following transport equations are derived to determine k and ε:

$$\frac{\partial}{\partial t}k + \vec{U}\mathrm{grad}k - \frac{\nu_T}{2}\left|\mathrm{grad}\vec{U} + \mathrm{grad}\vec{U}^T\right|^2 - \mathrm{div}(\nu_T \mathrm{grad}k) + \varepsilon = 0, \tag{10.18}$$

$$\frac{\partial}{\partial t}\varepsilon + \vec{U}\mathrm{grad}\varepsilon - \frac{c_1}{2}k\left|\mathrm{grad}\vec{U} + \mathrm{grad}\vec{U}^T\right|^2 - \mathrm{div}\left(\frac{c_\varepsilon}{c_\mu}\nu_T \mathrm{grad}\varepsilon\right) + c_2\frac{\varepsilon^2}{k} = 0. \tag{10.19}$$

The constants $c_\mu, c_\varepsilon, c_1$, and c_2 occurring here are usually chosen empirically in such a way that the model yields the correct results for three well-known flow cases (grid turbulence, turbulent shear flow, and turbulent flow over a flat plate). This calibration leads to

$$c_\mu = 0.09, \quad c_\varepsilon = 0.07, \quad c_1 = 0.126, \quad c_2 = 1.92. \tag{10.20}$$

A thorough derivation and analysis of the k-ε turbulence model may be found in [Mohammadi & Pironneau, 1993].

10.2.4 Boundary Conditions for the k-ε Model

Before we can solve the system of partial differential equations consisting of the momentum equations (10.16), (10.17), the continuity equation (10.7b), and equations (10.18), (10.19) for the turbulent kinetic energy k and the dissipation rate ε, we need to formulate appropriate boundary conditions.

A natural choice would be to prescribe k and ε or their normal derivatives along the boundary. It is, however, often unclear how these functions are to be chosen and, moreover, one encounters another difficulty of a more fundamental nature: the k-ε turbulence model is itself *no longer valid* in the vicinity of solid walls. One reason for this lies in the low local Reynolds number due to the small velocities near the wall, whereas the k-ε model is valid only for higher Reynolds numbers. Another reason is that, close to solid walls, the turbulent variations and fluctuations normal to the wall are damped, thus violating another assumption underlying the k-ε model, namely, that of isotropic turbulence.

In turbulent flows, a thin boundary layer develops along solid walls such as pipe walls or obstacle boundaries, which usually consists of three basic regions. In the immediate wall vicinity the flow velocity is very small due to the no-slip condition at the wall. In this region the flow in the boundary layer is laminar. Outside of this region there follows a transition zone in which the flow becomes locally unstable. This region exhibits a logarithmic velocity profile. Still further from the wall lies the turbulent region.

So-called *wall functions* describing the flow behavior near the wall are determined from approximate calculations of the flow profile in the wall region as well as empirically, from experimental data. These allow the laminar and logarithmic regions to be eliminated and to instead be modeled in the boundary conditions by way of the wall functions. The flow is then discretized and computed only in the turbulent region. This involves imposing the following boundary conditions (cf. [Mohammadi & Pironneau, 1993], [Viollet, 1981]) on the boundary Γ_{turb} of the turbulent region, which lies at a distance δ from the nearest wall of the actual flow domain,

$$k = (u^*)^2 c_\mu^{-1/2}, \qquad \varepsilon = \frac{(u^*)^3}{0.41\delta}, \tag{10.21}$$

$$\Phi_n = 0, \qquad \frac{\partial \Phi_t}{\partial n} = g(\Phi_t), \tag{10.22}$$

in which the reference values[84] u^* and δ^* are given by

$$u^* := \sqrt{\nu \left.\frac{\partial \Phi_t}{\partial n}\right|_\Gamma} \quad \text{and} \quad \delta^* := \frac{\nu}{u^*}.$$

This requires δ to lie in the logarithmic region $[20\delta^*, 100\delta^*]$. The wall function Φ is given implicitly by

$$\Phi_t \left| \nu \frac{\partial \Phi_t}{\partial n} \right|^{-1/2} = 2.44 \log \left(\frac{\delta}{\sqrt{\nu}} \left| \frac{\partial \Phi_t}{\partial n} \right|^{1/2} \right) + 5.5.$$

Additional problems arise for separating flows. In particular, it is unclear how regions of reversed flow are to be treated. Moreover, these boundary conditions are not easy to implement.

[84] As in Section 2.1, Φ_n denotes the normal and Φ_t the tangential component of the mean velocity $\vec{U} = (U, V)$.

One therefore attempts to adjust the coefficients of the original k-ε model in such a way that damping effects near the wall or boundary layer, respectively, are accounted for. This is the strategy followed in the *low-Reynolds-number models.* Here the constants c_μ, c_1, and c_2 in the k-ε model are multiplied by functions f_μ, f_1, and f_2 [Hanjalic & Launder, 1976]. These modifying functions satisfy $0 < f_\mu \leq 1$, $f_1 \geq 1$, and $0 < f_2 \leq 1$. They depend on two *local* Reynolds numbers

$$R_t := \frac{k^2}{\nu\varepsilon} \quad \text{and} \quad R_\delta :- \sqrt{k}\,\frac{\delta}{\nu},$$

in which δ denotes the distance to the nearest wall.

In [Mohammadi & Pironneau, 1993], following [Lam & Bremhorst, 1981], the definitions

$$\begin{aligned} f_\mu &:= (1-\exp(-0.0165R_\delta))^2\left(1+\frac{20.5}{R_t}\right), \\ f_1 &:= 1+\left(\frac{0.05}{f_\mu}\right)^3, \quad f_2 := 1-\exp(-R_t^2) \end{aligned} \tag{10.23}$$

are used. As boundary conditions for $\vec{U}$, k, and ε, one uses

$$\vec{U}=0, \qquad k=0, \qquad \frac{\partial\varepsilon}{\partial n}=0$$

along solid walls,

$$\frac{\partial\vec{U}}{\partial n}=0, \qquad \frac{\partial k}{\partial n}=0, \qquad \frac{\partial\varepsilon}{\partial n}=0$$

at outflow boundaries, and

$$\vec{U}=\vec{U}_{\text{in}}, \qquad k=k_{\text{in}}, \qquad \varepsilon=\varepsilon_{\text{in}}$$

along inflow boundaries with functions $\vec{U}_{\text{in}}$, k_{in}, and ε_{in} depending on the specific problem.

A survey on further low-Reynolds-number models can be found in [Patel et al., 1989] and [Rodi, 1994]. A deficiency of all these models is that they require relatively fine resolution near the wall to produce good results, since ε in particular possesses very steep gradients in the boundary layer next to the wall.

For this reason, so-called *two-layer models* are currently being developed in which the k-ε equations are used at a sufficiently large distance from the wall, while the viscous layer at the wall is described by a single-equation model in which the length scales are explicitly given [Mohammadi, 1992].

10.2.5 Overview of Other Turbulence Models

Reynolds stress models constitute another line of development, in which a separate transport equation is set up for each component of the Reynolds stress tensor.[85] These methods can be traced back to the work of [Launder et al., 1975].

[85]Since the tensor is symmetric, only the elements on or above the diagonal are required. This makes three components in the two-dimensional case and six for three dimensions.

Subsequent work is contained in [Gibson & Launder, 1978], [Hanjalic & Launder, 1976], [Jakirlic & Hanjalic, 1994], [Launder, 1990], and [Leschziner, 1989]. A recent survey of Reynolds stress models is given in [Rodi, 1994]. Despite their successful application in practice, the Reynolds stress models suffer from shortcomings in that they fail to satisfy what is known as the realizability condition [Schumann, 1977], and they also require more effort compared to low-Reynolds-number models. Further improvements of these models (*pressure-stress models, diffusion models*) are the objects of current research.

Further approaches to turbulence modeling are *nonlinear k-ε models, algebraic stress models*, and *renormalization group models*. Besides these statistically oriented methods, there are also fine-structure models based on the *large-eddy-simulation* idea. Smagorinsky's approach, as well as the current *dynamic-subgrid scale techniques*, also belongs in this class. Finally, *wavelet methods* have met with great interest in recent times as well, at least for the analysis of turbulent flows [Farge, 1992], [Meneveau, 1991], and may lead to new turbulence models. An (incomplete) overview of the various approaches to simulating turbulence is given in Figure 10.3.

10.3 Discretization of the k-ε Model

The discretization of the equations of the k-ε model proceeds similarly to that of the energy and momentum equations in the foregoing chapters. We will evaluate the turbulent kinetic energy k and its dissipation rate ε at the cell centers as we have done with pressure and temperature (cf. [Bader, 1995]). We further restrict ourselves to constant body forces $\vec{g}$ and use the low-Reynolds-number model described on page 163.

Discretization of the Reynolds Equations

In the same manner as in (3.34), (3.35) and (9.27), (9.28), respectively, we obtain the discrete version of the Reynolds equations (10.16) and (10.17):

$$U_{i,j}^{(n+1)} = \hat{F}_{i,j}^{(n)} - \frac{\delta t}{\delta x}\left(P_{i+1,j}^{(n+1)} - P_{i,j}^{(n+1)}\right), \qquad i = 1,\ldots,i_{\max}-1, \quad j = 1,\ldots,j_{\max}, \tag{10.24}$$

$$V_{i,j}^{(n+1)} = \hat{G}_{i,j}^{(n)} - \frac{\delta t}{\delta y}\left(P_{i,j+1}^{(n+1)} - P_{i,j}^{(n+1)}\right), \qquad i = 1,\ldots,i_{\max}, \quad j = 1,\ldots,j_{\max}-1, \tag{10.25}$$

which we evaluate at the midpoints of the vertical and horizontal edges as before. The new quantities $\hat{F}_{i,j}$ and $\hat{G}_{i,j}$ are defined as

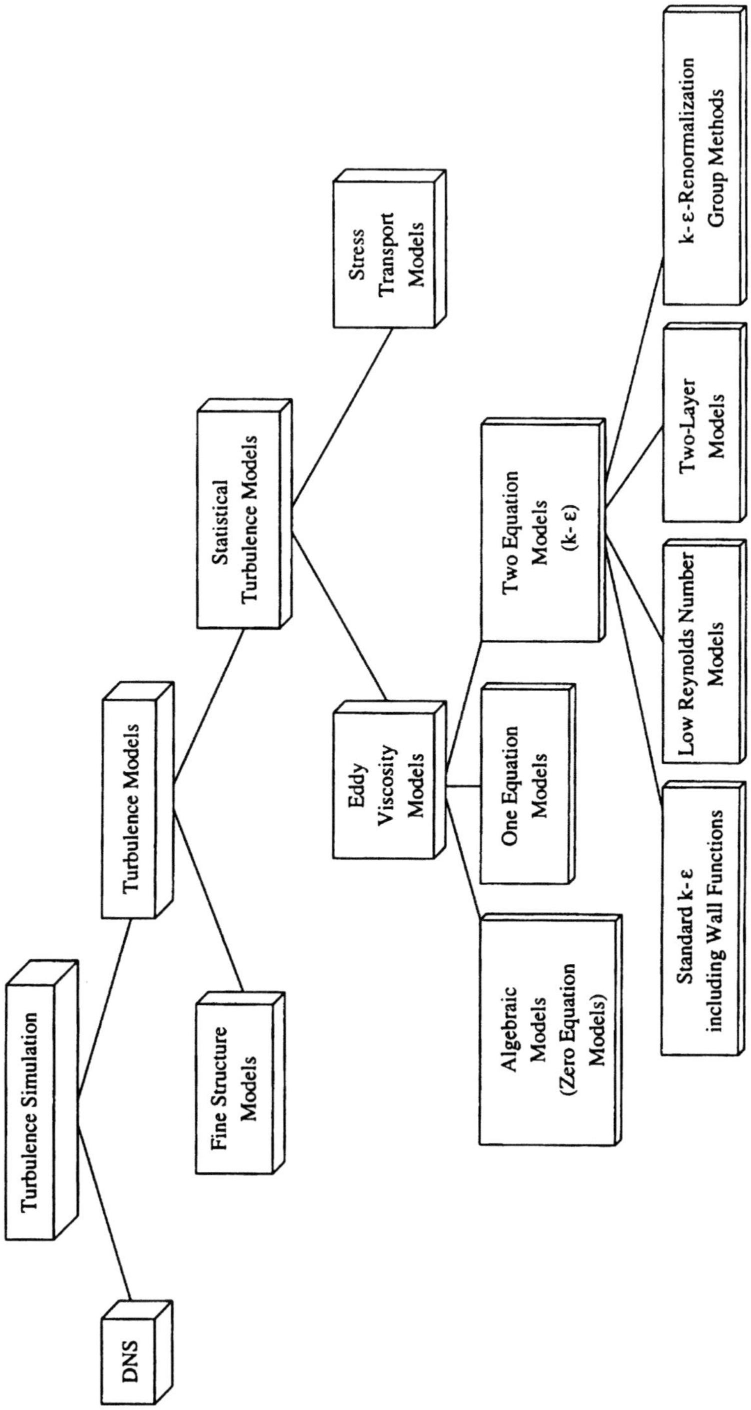

FIG. 10.3. *An overview of turbulence simulation.*

$$
\begin{aligned}
\hat{F}_{i,j} &:= U_{i,j} + \delta t \Bigg(2 \left[\frac{\partial}{\partial x} \left(\nu^* \frac{\partial U}{\partial x} \right) \right]_{i,j} + \left[\frac{\partial}{\partial y} \left(\nu^* \left(\frac{\partial U}{\partial y} + \frac{\partial V}{\partial x} \right) \right) \right]_{i,j} \\
&\qquad - \frac{2}{3} \left[\frac{\partial k}{\partial x} \right]_{i,j} - \left[\frac{\partial (U^2)}{\partial x} \right]_{i,j} - \left[\frac{\partial (UV)}{\partial y} \right]_{i,j} + g_x \Bigg), \\
&\qquad i = 1, \ldots, i_{\max} - 1, \quad j = 1, \ldots, j_{\max}, \\
\hat{G}_{i,j} &:= V_{i,j} + \delta t \Bigg(\left[\frac{\partial}{\partial x} \left(\nu^* \left(\frac{\partial V}{\partial x} + \frac{\partial U}{\partial y} \right) \right) \right]_{i,j} + 2 \left[\frac{\partial}{\partial y} \left(\nu^* \frac{\partial V}{\partial y} \right) \right]_{i,j} \\
&\qquad - \frac{2}{3} \left[\frac{\partial k}{\partial y} \right]_{i,j} - \left[\frac{\partial (UV)}{\partial x} \right]_{i,j} - \left[\frac{\partial (V^2)}{\partial y} \right]_{i,j} + g_y \Bigg), \\
&\qquad i = 1, \ldots, i_{\max}, \quad j = 1, \ldots, j_{\max} - 1 .
\end{aligned}
\tag{10.27}
$$

Here the values of U and V are those at time t_n, while the values of ε (and hence ν^*) are those at time t_{n+1}.

Besides the gradient of k, which is approximated in the same way as the gradient of p, we now must also discretize the new term

$$
\operatorname{div}(\nu^* (\operatorname{grad}\vec{U} + \operatorname{grad}\vec{U}^T)),
$$

in which the total viscosity is now no longer constant. Introducing the notation[86]

$$
\begin{aligned}
&[\nu_T]_{i,j} := c_\mu [f_\mu]_{i,j} \frac{k_{i,j}^2}{\epsilon_{i,j}}, \qquad \nu_{i,j}^* := \nu + [\nu_T]_{i,j}, \\
&\nu^*_{i+\frac{1}{2},j+\frac{1}{2}} := \frac{\nu^*_{i+1,j+1} + \nu^*_{i,j+1} + \nu^*_{i+1,j} + \nu^*_{i,j}}{4},
\end{aligned}
\tag{10.28}
$$

this is achieved by

$$
\left[\frac{\partial}{\partial x} \left(\nu^* \frac{\partial U}{\partial x} \right) \right]_{i,j} := \frac{1}{(\delta x)^2} \left(\nu^*_{i+1,j} (U_{i+1,j} - U_{i,j}) - \nu^*_{i,j} (U_{i,j} - U_{i-1,j}) \right),
$$

$$
\left[\frac{\partial}{\partial y} \left(\nu^* \frac{\partial V}{\partial y} \right) \right]_{i,j} := \frac{1}{(\delta y)^2} \left(\nu^*_{i,j+1} (V_{i,j+1} - V_{i,j}) - \nu^*_{i,j} (V_{i,j} - V_{i,j-1}) \right),
$$

$$
\begin{aligned}
\left[\frac{\partial}{\partial x} \left(\nu^* \left(\frac{\partial V}{\partial x} + \frac{\partial U}{\partial y} \right) \right) \right]_{i,j} := \frac{1}{\delta x} \Bigg(&\nu^*_{i+\frac{1}{2},j+\frac{1}{2}} \left(\frac{V_{i+1,j} - V_{i,j}}{\delta x} + \frac{U_{i,j+1} - U_{i,j}}{\delta y} \right) \\
&- \nu^*_{i-\frac{1}{2},j+\frac{1}{2}} \left(\frac{V_{i,j} - V_{i-1,j}}{\delta x} + \frac{U_{i-1,j+1} - U_{i-1,j}}{\delta y} \right) \Bigg).
\end{aligned}
$$

[86]The notation $[f_\mu]_{i,j}$, $[f_1]_{i,j}$, and $[f_2]_{i,j}$ means that the discrete values $k_{i,j}$ and $\varepsilon_{i,j}$ are to be inserted in the formulas (10.23) for f_μ, f_1, and f_2.

$$\left[\frac{\partial}{\partial y}\left(\nu^*\left(\frac{\partial U}{\partial y}+\frac{\partial V}{\partial x}\right)\right)\right]_{i,j} := \frac{1}{\delta y}\left(\nu^*_{i+\frac{1}{2},j+\frac{1}{2}}\left(\frac{U_{i,j+1}-U_{i,j}}{\delta y}+\frac{V_{i+1,j}-V_{i,j}}{\delta x}\right)\right.$$
$$\left. -\nu^*_{i+\frac{1}{2},j-\frac{1}{2}}\left(\frac{U_{i,j}-U_{i,j-1}}{\delta y}+\frac{V_{i+1,j-1}-V_{i,j-1}}{\delta x}\right)\right), \quad (10.30)$$

Discretization of the Transport Equations for k and ε

As in the case of the energy equation in (9.20), the transport equations of k (10.18) and ε (10.19) are discretized at the cell centers:

$$\begin{aligned} k_{i,j}^{(n+1)} = k_{i,j}^{(n)} + \delta t \Bigg(& \left[\frac{\partial}{\partial x}\left(\nu_T\frac{\partial k}{\partial x}\right)\right]_{i,j} + \left[\frac{\partial}{\partial y}\left(\nu_T\frac{\partial k}{\partial y}\right)\right]_{i,j} \\ & - \left[\frac{\partial(Uk)}{\partial x}\right]_{i,j} - \left[\frac{\partial(Vk)}{\partial y}\right]_{i,j} \\ & + \frac{1}{2}[\nu_T]_{i,j}\left[|\mathrm{grad}\vec{U}+\mathrm{grad}\vec{U}^T|^2\right]_{i,j} - \varepsilon_{i,j}\Bigg), \end{aligned} \quad (10.31)$$

$$\begin{aligned} \varepsilon_{i,j}^{(n+1)} = \varepsilon_{i,j}^{(n)} + \delta t \Bigg(& \frac{c_\varepsilon}{c_\mu}\left[\frac{\partial}{\partial x}\left([f_\mu]_{i,j}\nu_T\frac{\partial \varepsilon}{\partial x}\right)\right]_{i,j} + \frac{c_\varepsilon}{c_\mu}\left[\frac{\partial}{\partial y}\left([f_\mu]_{i,j}\nu_T\frac{\partial \varepsilon}{\partial y}\right)\right]_{i,j} \\ & - \left[\frac{\partial(U\varepsilon)}{\partial x}\right]_{i,j} - \left[\frac{\partial(V\varepsilon)}{\partial y}\right]_{i,j} \\ & + \frac{c_1[f_1]_{i,j}}{2}k_{i,j}\left[|\mathrm{grad}\vec{U}+\mathrm{grad}\vec{U}^T|^2\right]_{i,j} - c_2[f_2]_{i,j}\frac{\varepsilon_{i,j}^2}{k_{i,j}}\Bigg). \end{aligned} \quad (10.32)$$

The discretization of most of these terms is already familiar from (3.34), (3.35), and (9.21) (replace T in the convective terms by k and ε, respectively). The following are new:

$$\left[\frac{\partial}{\partial x}\left(\nu_T\frac{\partial k}{\partial x}\right)\right]_{i,j} := \frac{1}{(\delta x)^2}\left([\nu_T]_{i+\frac{1}{2},j}(k_{i+1,j}-k_{i,j}) - [\nu_T]_{i-\frac{1}{2},j}(k_{i,j}-k_{i-1,j})\right),$$

$$\left[\frac{\partial}{\partial y}\left(\nu_T\frac{\partial k}{\partial y}\right)\right]_{i,j} := \frac{1}{(\delta y)^2}\left([\nu_T]_{i,j+\frac{1}{2}}(k_{i,j+1}-k_{i,j}) - [\nu_T]_{i,j-\frac{1}{2}}(k_{i,j}-k_{i,j-1})\right),$$

$$\left[|\mathrm{grad}\vec{U}+\mathrm{grad}\vec{U}^T|^2\right]_{i,j} := 4\left[\frac{\partial U}{\partial x}\right]_{i,j}^2 + 2\left(\left[\frac{\partial U}{\partial y}\right]_{i,j} + \left[\frac{\partial V}{\partial x}\right]_{i,j}\right)^2 + 4\left[\frac{\partial V}{\partial y}\right]_{i,j}^2,$$

$$\left[\frac{\partial U}{\partial x}\right]_{i,j} := \frac{U_{i,j}-U_{i-1,j}}{\delta x},$$

$$\left[\frac{\partial U}{\partial y}\right]_{i,j} := \frac{(U_{i,j+1}+U_{i-1,j+1})-(U_{i,j-1}+U_{i-1,j-1})}{4\delta y},$$

$$\left[\frac{\partial V}{\partial y}\right]_{i,j} := \frac{V_{i,j} - V_{i,j-1}}{\delta y},$$
$$\left[\frac{\partial V}{\partial x}\right]_{i,j} := \frac{(V_{i+1,j} + V_{i+1,j-1}) - (V_{i-1,j} + V_{i-1,j-1})}{4\delta x}. \tag{10.34}$$

Here the values $[\nu_T]_{i\pm\frac{1}{2},j}$ and $[\nu_T]_{i,j\pm\frac{1}{2}}$ of the turbulent eddy viscosity at the edge midpoints are determined by averaging the values in the neighboring cell centers. U, V, k, and ε on the right-hand side of (10.31) and (10.32) are all evaluated at time t_n.

Boundary Conditions for k and ε

For the low-Reynolds-number model, the boundary conditions for k and ε are simply Dirichlet and Neumann boundary conditions discretized in the same manner as the corresponding boundary conditions for the temperature in (9.22) and (9.23) on page 134.

The Extended Algorithm

The sequence of computations to be performed is summarized once more in Algorithm 5.

```
Set t := 0, n := 0
Assign initial values to U, V, P, k, ε
While t < t_end
    Set the boundary values for U, V, k, and ε
    Compute k^(n+1) and ε^(n+1) according to (10.31) and (10.32)
    Compute F̂^(n) and Ĝ^(n) according to (10.27)
    Compute the right-hand side of the pressure equation (3.38)
    Set it := 0
    While it < it_max and ||r^it|| > eps (resp., ||r^it|| > eps||P^0||)
        Perform an SOR cycle according to (3.44)
        Compute the residual norm of the pressure equation ||r^it||
        it := it + 1
    Compute U^(n+1) and V^(n+1) according to (10.24) and (10.25)
    t := t + δt
    n := n + 1
```

Algorithm 5. Extension to include the k-ε model.

10.4 Implementation

To implement the k-ε model, the basic program from Chapter 3 must be modified as follows.

- The quantities

  `REAL KAI,EPI` initial values of k and ε

- and the arrays

```
REAL **KA          turbulent kinetic energy k,
REAL **EP          dissipation rate ε
```

of dimension `[0,imax+1]x[0,jmax+1]` must be defined and allocated. In addition, the following functions must be introduced or modified.

1. `void COMP_KAEP(U,V,KA,EP,FLAG,imax,jmax,delt,delx,dely,GX,GY,Re, gamma)`: Computation of $k^{(n+1)}$ and $\varepsilon^{(n+1)}$ according to (10.31) and (10.32). This function is called in `main` after setting the boundary values. It may be incorporated into the module `uvp.c`.
2. The function `READ_PARAMETER` must be extended to read the initial values `KAI` and `EPI` for k and ε.
3. The function `COMP_FG` must be modified in accordance with formula (10.27).
4. In function `INIT_UVP`, k and ε must be initialized as well.
5. In `SETBCOND` and `SETSPECBCOND`, the boundary conditions for `KA` and `EP` are to be included.
6. The function `OUTPUTVEC` must be augmented by the output of `KA` and `EP`.
7. The main program from page 40 must be extended accordingly.

10.5 Numerical Results

For the flow over a backward-facing step at Reynolds number $Re = 20000$ (see Section 5.2), one obtains the results shown in Figures 10.4, 10.5, and 10.6 when using the low-Reynolds-number model. The following simplified inflow functions were used at the left boundary:

$$U_{\text{in}} = 1, \qquad V_{\text{in}} = 0, \qquad k_{\text{in}} = 0.003\, U_{\text{in}}^2, \qquad \varepsilon_{\text{in}} = c_\mu \frac{k_{\text{in}}^{3/2}}{0.03\, b}$$

with a channel width of b (cf. also [Chang et al., 1991]). For a more realistic simulation, one would have to use a turbulent inflow velocity profile which could, for example, be obtained from experimental data. Moreover, the resolution near the wall should be finer, which could be achieved by adaptive techniques. In our simulation, though, we only used a grid consisting of 120×60 cells of equal size. Compared with the results of the laminar simulation in Section 5.2, Figure 5.9, the recirculation area behind the step is much smaller, and the upper eddy obtained in the laminar calculation for high Reynolds numbers completely disappears in the turbulent simulation; see Figure 10.5. Moreover, the profile of the mean velocity in x-direction U is flattened in contrast to the parabolic profile obtained without turbulence models in Figure 5.10. Thus, turbulence has substantial influence on the flow properties.

Results of laboratory experiments can be found in [Driver & Seegmüller, 1985] and [Kim et al., 1980]. Numerical results are discussed in, among others, [Schmitt, 1988].

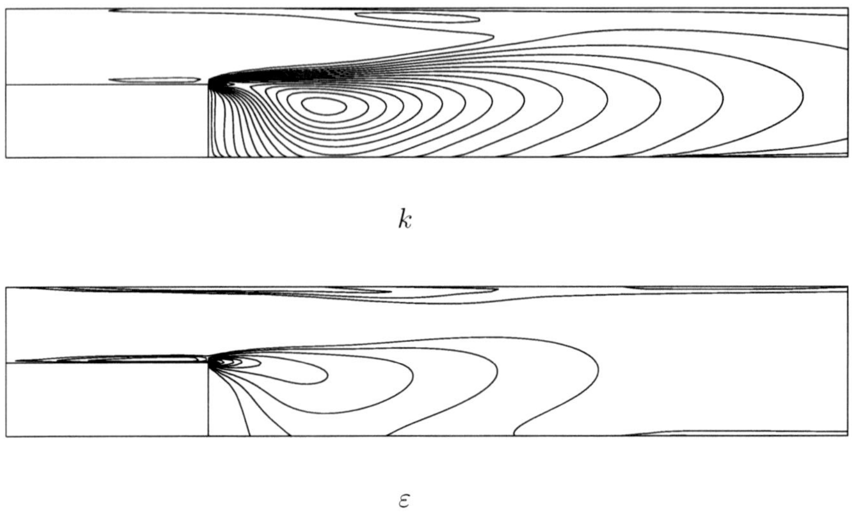

FIG. 10.4. *Turbulent flow over a backward-facing step at* $Re = 20000$, k *and* ε *contours.*

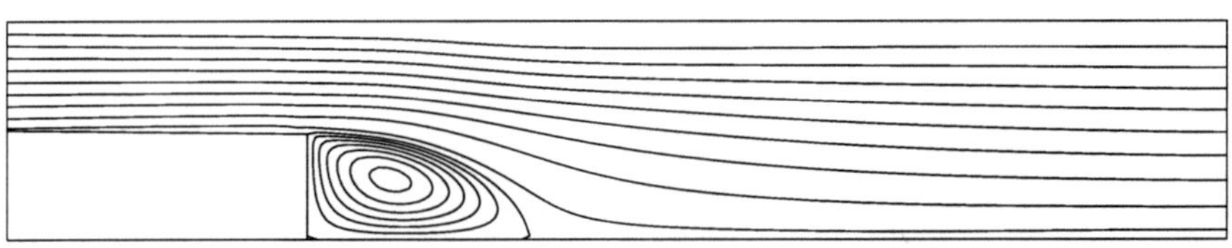

FIG. 10.5. *Turbulent flow over a backward-facing step at* $Re = 20000$, *streamlines.*

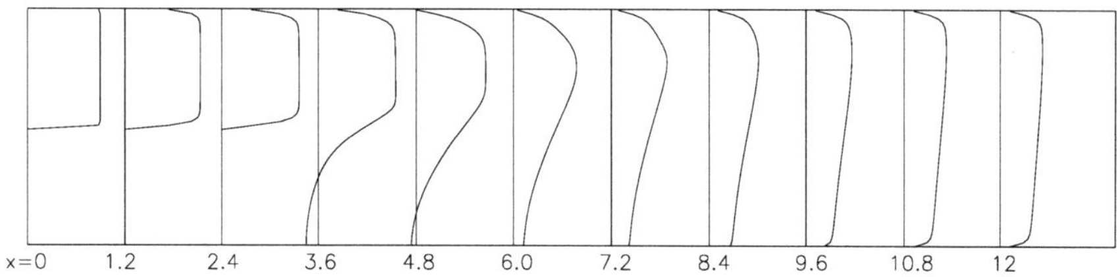

U

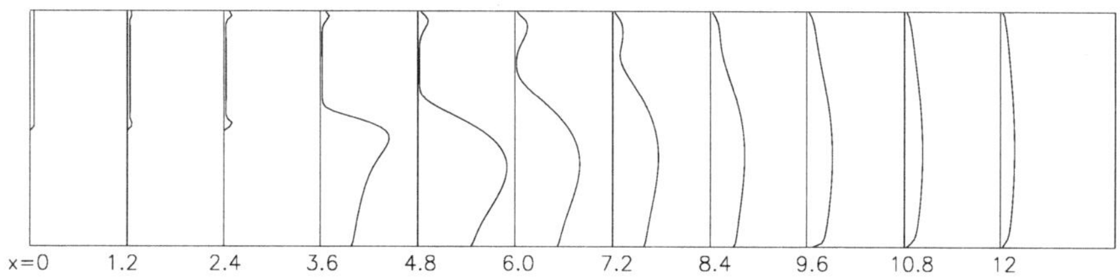

k

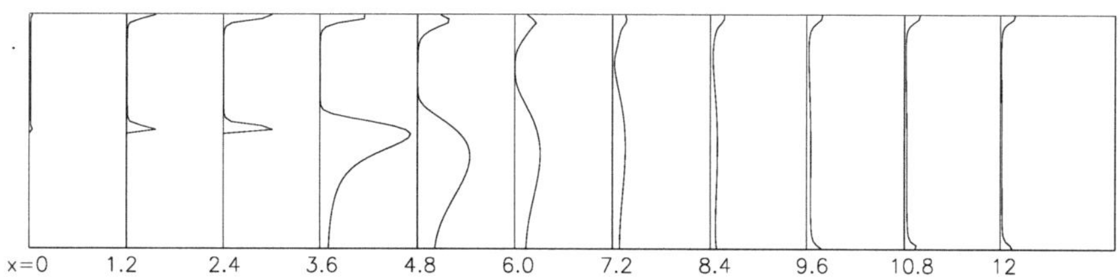

ε

FIG. 10.6. *Turbulent flow over a backward-facing step at* $Re = 20000$*, vertical profiles of* U*,* k*, and* ε*.*

Chapter 11

Extension to Three Dimensions

We conclude this book with a small excursion into the world of three-dimensional simulation, which of course captures the real world much better than the two-dimensional projections we have been considering so far. This more realistic simulation, however, comes with a price tag of considerably higher requirements in memory and CPU time, so that three-dimensional simulations can generally not be performed with as fine a resolution as two-dimensional ones. In the same way, the visualization of three-dimensional data sets on a computer screen or on paper is only possible using more refined techniques such as projections and cross-sections.

From the mathematical-numerical standpoint, though, we encounter no fundamental difficulties in extending the basic two-dimensional code we have so far put together to the three-dimensional case.

11.1 The Continuous Equations

In the three-dimensional case, the dimensionless Navier–Stokes equations (2.1) (page 11) and the temperature equation (9.19) (page 132), written in component form, are given as follows.

Momentum equations:

$$\frac{\partial u}{\partial t} + \frac{\partial p}{\partial x} = \frac{1}{Re}\left(\frac{\partial^2 u}{\partial x^2} + \frac{\partial^2 u}{\partial y^2} + \frac{\partial^2 u}{\partial z^2}\right) - \frac{\partial (u^2)}{\partial x} - \frac{\partial (uv)}{\partial y} - \frac{\partial (uw)}{\partial z} + g_x\,, \tag{11.1a}$$

$$\frac{\partial v}{\partial t} + \frac{\partial p}{\partial y} = \frac{1}{Re}\left(\frac{\partial^2 v}{\partial x^2} + \frac{\partial^2 v}{\partial y^2} + \frac{\partial^2 v}{\partial z^2}\right) - \frac{\partial (uv)}{\partial x} - \frac{\partial (v^2)}{\partial y} - \frac{\partial (vw)}{\partial z} + g_y\,, \tag{11.1b}$$

$$\frac{\partial w}{\partial t} + \frac{\partial p}{\partial z} = \frac{1}{Re}\left(\frac{\partial^2 w}{\partial x^2} + \frac{\partial^2 w}{\partial y^2} + \frac{\partial^2 w}{\partial z^2}\right) - \frac{\partial (uw)}{\partial x} - \frac{\partial (vw)}{\partial y} - \frac{\partial (w^2)}{\partial z} + g_z\,. \tag{11.1c}$$

Continuity equation:

$$\frac{\partial u}{\partial x} + \frac{\partial v}{\partial y} + \frac{\partial w}{\partial z} = 0\,. \tag{11.1d}$$

Energy equation:

$$\frac{\partial T}{\partial t} = \frac{1}{Re}\frac{1}{Pr}\left(\frac{\partial^2 T}{\partial x^2} + \frac{\partial^2 T}{\partial y^2} + \frac{\partial^2 T}{\partial z^2}\right) - \frac{\partial(uT)}{\partial x} - \frac{\partial(vT)}{\partial y} - \frac{\partial(wT)}{\partial z} + q'''. \quad (11.1e)$$

Here z denotes the third spatial coordinate, w the velocity component in the z-direction, and g_z the z-component of the body forces. As initial conditions at time $t = 0$, the velocities $u = u_0(x, y, z)$, $v = v_0(x, y, z)$, and $w = w_0(x, y, z)$ are required, and along the boundary of the domain Ω, which is now three-dimensional, one of the three following conditions must be prescribed for each of the three velocity components u, v, and w for all time. If we consider only domains whose boundary surfaces run parallel to the coordinate planes, then, at each of the boundary faces, one velocity component (φ_n) normal to the wall as well as two tangential components (φ_{t_1} and φ_{t_2}) need to be determined. Depending on the local orientation, each of φ_n, φ_{t_1}, and φ_{t_2} correspond to one of u, v, or w. Along a wall parallel to the x-z-plane, for example, one would have

$$\varphi_n := v, \qquad \varphi_{t_1} := u, \qquad \varphi_{t_2} := w.$$

The same boundary conditions as in two dimensions arise here as well.

1. No-slip:	$\varphi_n(x,y,z) = 0$,	$\varphi_{t_1}(x,y,z) = 0$,	$\varphi_{t_2}(x,y,z) = 0$.
2. Free-slip:	$\varphi_n(x,y,z) = 0$,	$\partial\varphi_{t_1}(x,y,z)/\partial n = 0$,	$\partial\varphi_{t_2}(x,y,z)/\partial n = 0$.
3. Inflow:	$\varphi_n(x,y,z) = \varphi_n^0$,	$\varphi_{t_1}(x,y,z) = \varphi_{t_1}^0$,	$\varphi_{t_2}(x,y,z) = \varphi_{t_2}^0$, $\varphi_n^0, \varphi_{t_1}^0, \varphi_{t_2}^0$ given.
4. Outflow:	$\partial\varphi_n(x,y,z)/\partial n = 0$,	$\partial\varphi_{t_1}(x,y,z)/\partial n = 0$,	$\partial\varphi_{t_2}(x,y,z)/\partial n = 0$.
5. Periodic (x-direction):	$\varphi_n(x,y,z) = \varphi_n(x+a,y,z)$,	$\varphi_{t_1}(x,y,z) = \varphi_{t_1}(x+a,y,z)$,	$\varphi_{t_2}(x,y,z) = \varphi_{t_2}(x+a,y,z)$.

11.2 Discretization and Algorithm

The base domain

$$\mathcal{G} := [0, a] \times [0, b] \times [0, c] \subset \mathbb{R}^3,$$

which is now a three-dimensional cuboid, is subdivided along the coordinate directions into $i_{\max}\, j_{\max}\, k_{\max}$ small cuboids with edge lengths δx, δy, and $\delta z := c/k_{\max}$. Pressure and temperature are localized at the centers of the subcuboids in a three-dimensional staggered grid, while the velocities u, v, and w are evaluated in the centers of the corresponding boundary faces (see Figure 11.1). Each discrete value is now assigned three indices i, j, and k, i and j being the indices in the x- and y-directions as before, and k the index in the z-direction.

The individual terms in the Navier–Stokes equations are now discretized analogously to the two-dimensional case (see (3.16) on page 28, (3.19a), (3.19c) on page 29, and (9.21) on page 133).

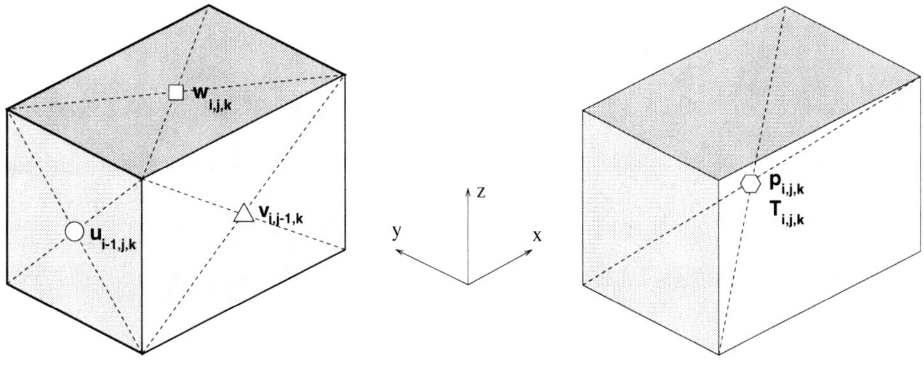

FIG. 11.1. *Grid cell of a staggered grid in three dimensions: velocities in the centers of the boundary faces (left), pressure and temperature at the cell center (right).*

In the time-stepping loop we obtain the time-discrete momentum equations

$$\begin{aligned} u^{(n+1)} &= F^{(n)} - \delta t \frac{\partial p^{(n+1)}}{\partial x}, \\ v^{(n+1)} &= G^{(n)} - \delta t \frac{\partial p^{(n+1)}}{\partial y}, \\ w^{(n+1)} &= H^{(n)} - \delta t \frac{\partial p^{(n+1)}}{\partial z}, \end{aligned} \tag{11.2}$$

in which, analogously to equation (3.29) (page 33), F, G, and H now contain the spatial derivatives of the velocities from equations (11.1a–c) as well as the velocities u, v, and w themselves, all of which are evaluated at time t_n. Together with the continuity equation (11.1d), this results in the three-dimensional Poisson equation

$$\frac{\partial^2 p^{(n+1)}}{\partial x^2} + \frac{\partial^2 p^{(n+1)}}{\partial y^2} + \frac{\partial^2 p^{(n+1)}}{\partial z^2} = \frac{1}{\delta t}\left(\frac{\partial F^{(n)}}{\partial x} + \frac{\partial G^{(n)}}{\partial y} + \frac{\partial H^{(n)}}{\partial z}\right) \tag{11.3}$$

for the pressure $p^{(n+1)}$. The latter is again solved using an SOR iteration, which now involves three nested loops over i, j, and k.

The stability conditions in three dimensions read (cf. (3.49) on page 39 and (9.24) on page 134)

$$\begin{gathered} \frac{2\,\delta t}{Re} < \left(\frac{1}{\delta x^2} + \frac{1}{\delta y^2} + \frac{1}{\delta z^2}\right)^{-1}, \quad \frac{2\,\delta t}{Re\,Pr} < \left(\frac{1}{\delta x^2} + \frac{1}{\delta y^2} + \frac{1}{\delta z^2}\right)^{-1}, \\ |u_{\max}|\,\delta t < \delta x, \quad |v_{\max}|\,\delta t < \delta y, \quad |w_{\max}|\,\delta t < \delta z. \end{gathered} \tag{11.4}$$

Otherwise, the solution procedure is the same as in the two-dimensional case (compare Section 3.2).

11.3 Extensions and Modifications

General Geometries

The treatment of general geometries as in Section 3.4 now requires distinguishing three different types of boundary cells in order to describe obstacles in the base domain $\mathcal{G}$: surface cells which border only on one fluid cell, edge cells which border on two fluid cells, and, finally, corner cells which border on three fluid cells. When discretizing no-slip conditions, velocities located on faces separating obstacle and fluid cells are set to zero. In contrast, velocities between two boundary cells are set to the negative value of the adjacent velocity value in the fluid region, resulting in a zero tangential velocity component on the boundary (cf. (3.51) and (3.52) on page 47).

The Treatment of Free Boundary Value Problems

Developing a code to solve free boundary value problems in three dimensions, however, requires substantial effort. While only 15 different surface cell types had to be distinguished in two dimensions (cf. Chapter 6), this number now rises to 63 in three dimensions; we will not further elaborate upon this here. In addition, the memory and CPU requirements for managing the particles are rather high, since each cell should now contain 27 or even 64 particles in the initial configuration rather than 9 or 16, as in the two-dimensional case. If the particle density is chosen too low, then it may happen that all particles of one cell leave the cell in the course of the computation, leading to this cell being mistakenly classified as an empty cell. For this reason, efficient methods in three dimensions mandate other techniques such as the *volume-of-fluid* method or methods which track particles only along the free boundary or in cells close to the free boundary (see also the end of Section 6.1).

On Parallelization

Of course, the benefits of parallelization can be significantly greater in the three-dimensional case than for the two-dimensional one, since the data size is now substantially larger if the base domain is to be well resolved by the numerical grid. The extension of the two-dimensional code to the three-dimensional case is straightforward and presents no major difficulties.

In contrast with the parallel algorithm in Chapter 8, the three-dimensional case requires exchanging data on all faces among neighboring processors rather than just data along edges. The data exchange must now consist of six rather than four stages, i.e., to the left, right, up, down, front, and rear (cf. page 114).

11.4 Examples of Three-Dimensional Simulations

In conclusion, we present some examples of three-dimensional simulations computed using an extended version of the code. Since the visualization of data in three dimensions is substantially more complex compared to two dimensions, software allowing for the color rendering of the computed results and the generation of perspective views becomes crucial. In this context we also refer to [Keller & Keller, 1993], which contains an extensive collection of examples of three-dimensional scientific visualization.

The examples shown are no longer analyzed in detail; their purpose is rather to capture the reader's interest and to provide the motivation for performing independent numerical investigations using the program. We consider two simple model problems, various applications from environmental sciences, engineering, and architecture, two examples of free boundary value problems, and convective three-dimensional flows driven by temperature gradients. The environmental, engineering, and architectural applications, in particular, are not always entirely realistic, since the grid resolution is often too coarse, the Reynolds numbers used are too low (particularly for air flows), and no turbulence model is included. These cases warrant moving to a large parallel computer as well as adding a suitable turbulence model to the code.

11.4.1 Model Problems

Flow over a Backward-Facing Step

To visualize the three-dimensional simulation of the flow over a backward-facing step (Reynolds number $Re = 10$, resolution $120 \times 25 \times 25$), Figures 11.2 and 11.3 show the color-coded velocity magnitudes in a section along the central vertical plane lying parallel to the main flow direction. Red denotes high velocities; blue, low velocities. In addition, the velocities' direction and magnitude can be inferred from the arrows.

The color coding of the velocity magnitude along the central plane reveals the development of an eddy region just beyond the step, as was the case with the two-dimensional simulation (cf. Chapter 5, p. 76). This eddy region is shown magnified in Figure 11.3.

Figure 11.4 shows several streamlines inside the eddy behind the step as well as an isosurface of the velocity component in the principal flow direction. This isosurface is further colorized with a color coding representing the horizontal velocity component normal to the principal flow direction. This coloring shows the fluid in the eddy being forced to the inside by the lateral walls. The shape of the streamlines further illustrates this observation.

Flow Around an Obstacle

The simulation of the flow around a fixed rounded obstacle on the floor of a channel yields an unsteady flow behavior similar to that observed in the two-dimensional simulation of the flow past an obstacle (cf. Section 5.3). Figures 11.5

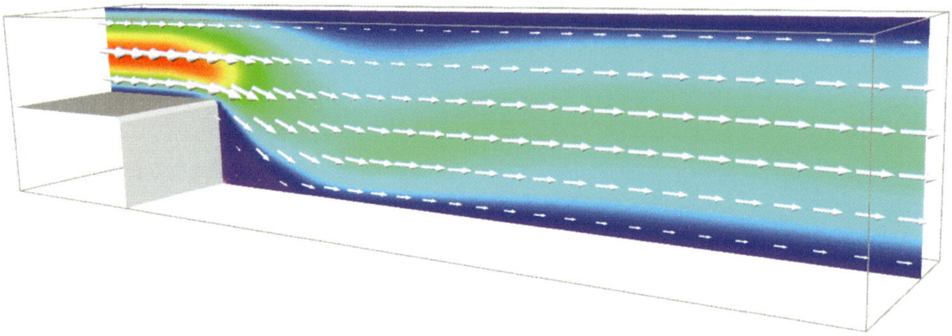

FIG. 11.2. *Flow over a backward-facing step.*

FIG. 11.3. *Flow over a backward-facing step, magnification of the eddy behind the step.*

and 11.6 show the flow (Reynolds number $Re = 7576$, resolution $120 \times 25 \times 25$, boundary conditions: inflow on the left, outflow on the right, no-slip otherwise) at a fixed time level. Of these, the streamline images of Figure 11.5 nicely reveal the three-dimensional vortical patterns forming behind the obstacle. The two images show streamlines starting at various points in the flow at a fixed time instance.

Figure 11.6 shows the velocity field along two horizontal sections, in which the vector field is additionally color-coded to show the velocity magnitude. The eddy structure is also clearly visible here.

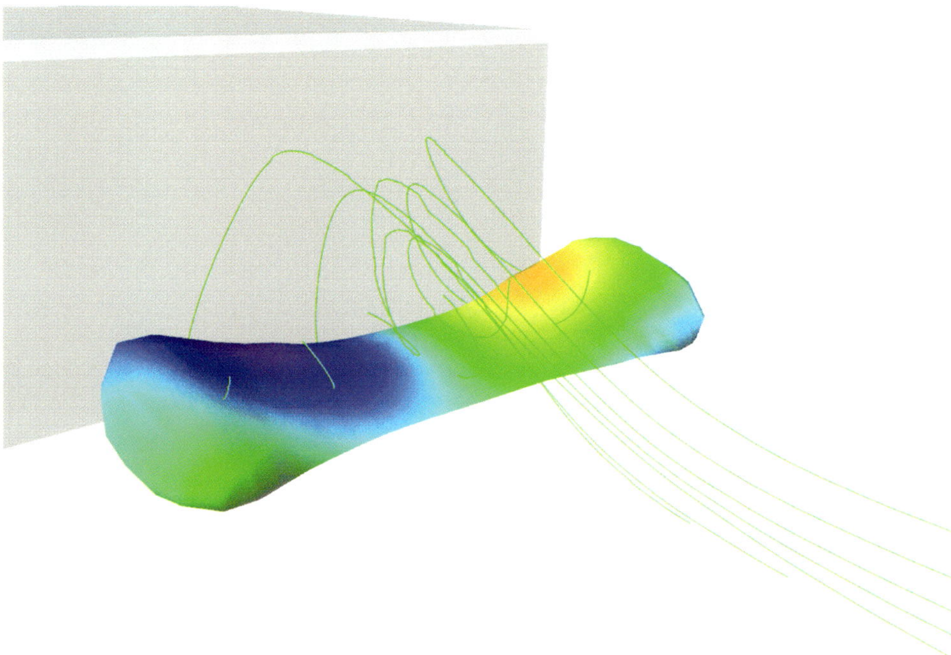

FIG. 11.4. *Flow over a backward-facing step, illustration of the three-dimensionality of the eddy structure.*

11.4.2 Examples from Environmental Sciences, Architecture, and Engineering

Flow of Air over Terrain

Geographers often represent topographical structures such as mountains, valleys, islands, etc. with the help of a *digital elevation model* (DEM) resulting from stereogrammetric measurements. This requires special processing of aerial or satellite images. In the simplest case, a topography is described by an elevation function over a uniform two-dimensional grid. Such a representation is well suited for our flag-array technique for the approximate description of complicated geometries. The generation of the flag array then merely consists of three nested loops (over `i`, `j`, and `k`), in which simple comparisons determine whether each cell lies in the fluid domain or in the obstacle domain. The accuracy with which the elevation function is represented is determined by the resolution of the three-dimensional simulation grid.

The flow shown in Figure 11.7 over the mountain region surrounding the Zugspitze, a mountain in the Bavarian Alps, is a first example of the flexibility of the flag-array technique for treating general geometries also in three dimensions. The

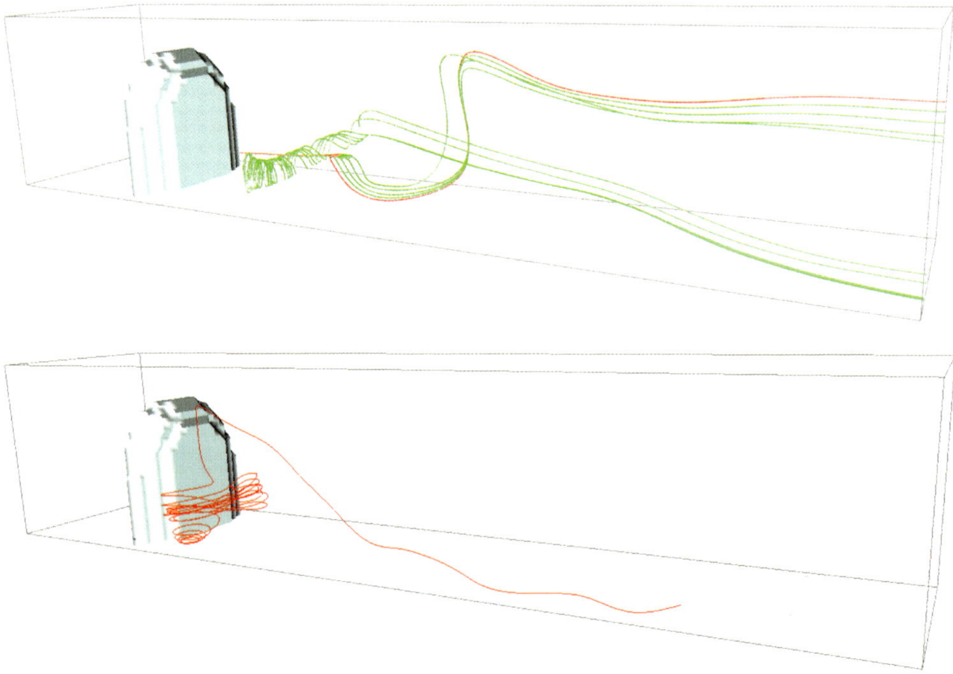

FIG. 11.5. *Streamlines of the flow past an obstacle.*

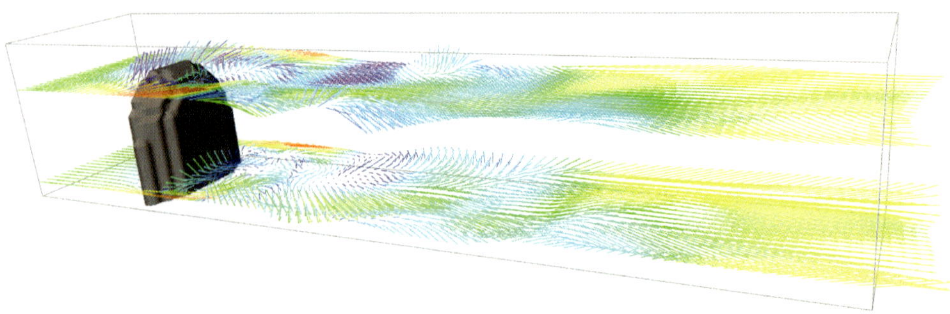

FIG. 11.6. *Velocity vectors in two horizontal sections; color coding shows velocity magnitude.*

velocity field is again represented using arrows of different lengths and colors. The color-coding of the mountain represents elevation.

A second example of air flowing across terrain is given in Figure 11.8. It shows the flow around the island of Jan Mayen in the North Atlantic, which measures roughly 60 km by 15 km. A unique feature of this island is its almost perfect volcanic cone Beerenberg, extending to 2300 m above sea level at its highest elevation. For the numerical calculation, a grid of $60\times29\times40$ cells was used, yielding only a very coarse resolution of the island's shape.

The visualization of the island in Figure 11.8 has been stretched in the z-

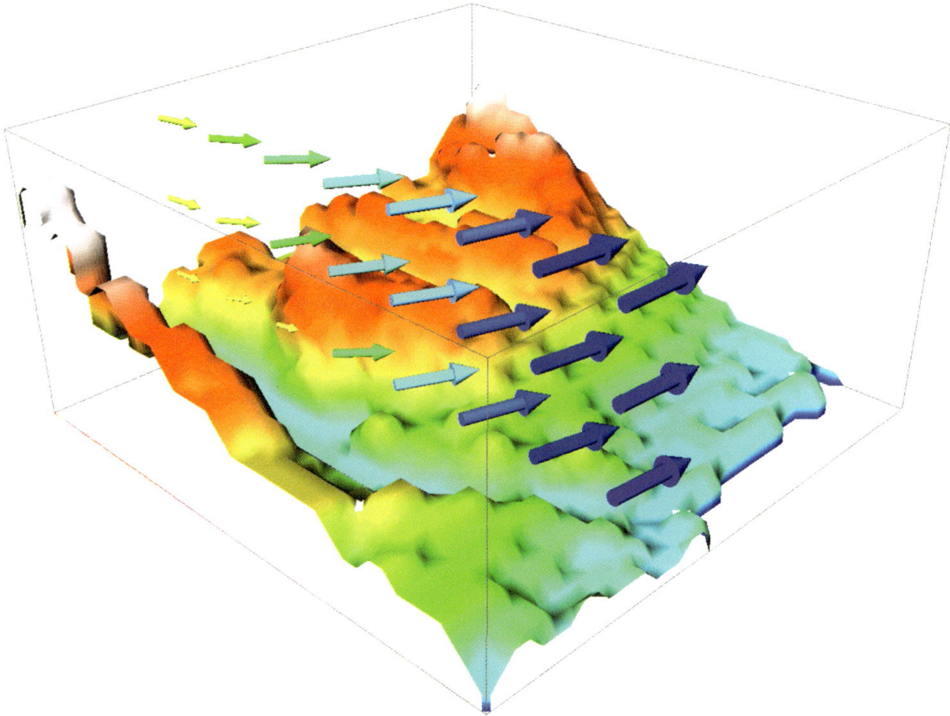

FIG. 11.7. *Flow of air over a mountain region.*

direction. The color-coding of the island represents elevation. The unsteady flow field around the island is illustrated at a fixed time by several streamlines as well as two horizontal sections showing color-coded velocity magnitudes. Red indicates high velocities; blue, low velocities. The color-coding in certain velocity ranges has been suppressed (by making them transparent) to prevent the sections from obscuring too much of the image. This representation illustrates the vorticity structure just in the lee of the island.

It is also interesting to observe the flow long after it has passed over the island; [Blum, 1993] contains a satellite image of a "typical cloud phenomenon in the North Atlantic" on page 13: "cold wind from the polar marine ice region is parted by the Norwegian island of Jan Mayen and forms a braid-like *Kármán vortex street* in its lee." A section of this satellite image is shown in Figure 11.9. It shows the Kármán vortex street and at the same time gives an impression of the large scale of this flow. We are familiar with this phenomenon from a two-dimensional simulation in Section 5.3; the similarity between the satellite image and Figure 5.14 on page 80 is apparent.

Flow Patterns around Tall Buildings

The flow patterns around buildings in a city are of interest for a number of reasons. We mention only the spreading of pollutants generated by automobile traffic or emitted from industrial smokestacks or the magnitude of wind stresses

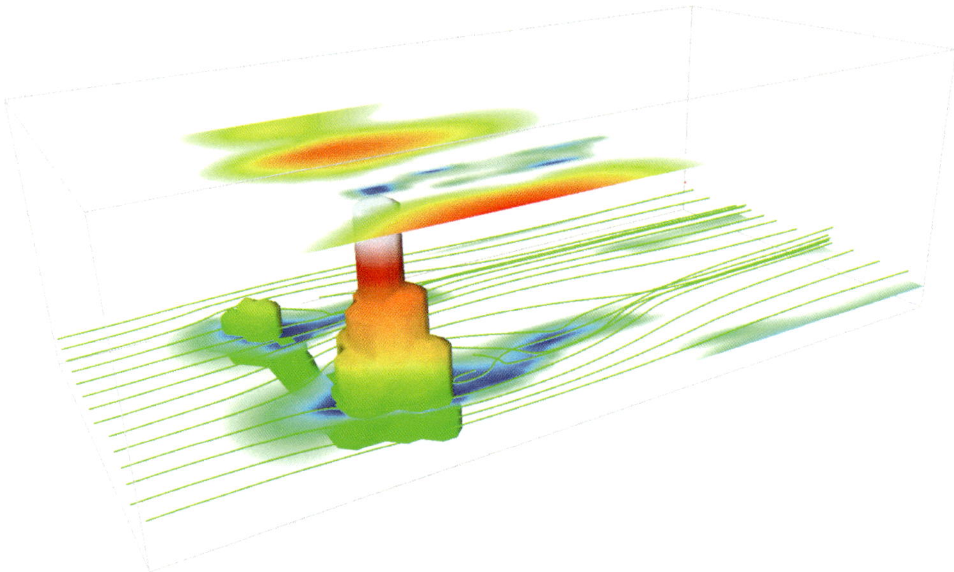

FIG. 11.8. *Flow of air over Jan Mayen Island in the North Atlantic.*

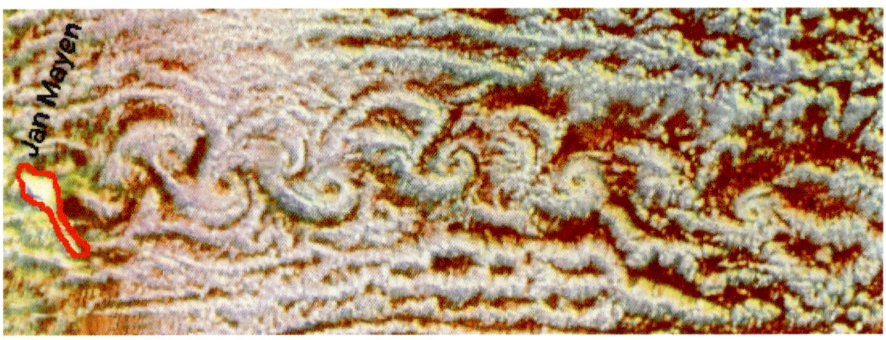

FIG. 11.9. *Cloud vortices in the wake of Jan Mayen Island in the North Atlantic, excerpt from a photograph taken by NOAA satellite.*

acting on tall structures. In these issues, city planners and architects may obtain design guidelines from numerical simulation.

Figure 11.10 shows the representation of a simplified model of a city containing one tall building at a grid resolution of $100 \times 50 \times 30$ cells. The free-stream velocity is given by the arrows along the left boundary, and the flow field is made visible by color-coded velocity magnitudes in two horizontal sections.

Room Ventilation

Besides the flow around buildings, the flow *inside* buildings is often the object of fluid engineering investigations. In this context, a common problem is the effec-

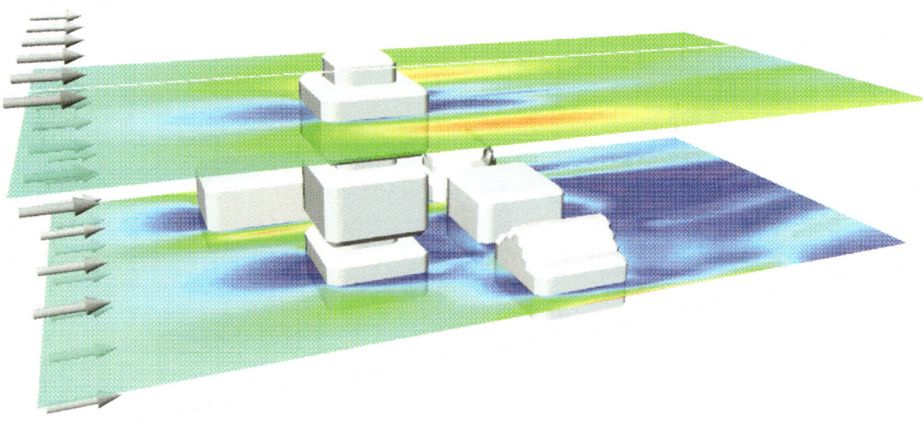

FIG. 11.10. *Flow pattern around tall buildings.*

tiveness of room ventilation provided by air conditioning systems in department stores, office buildings, etc. Figures 11.11 and 11.12 show the results of the numerical simulation of the ventilation of a model office floor. The obstacle structure here consists of three work spaces with desks separated by two walls reaching not quite up to the ceiling, two doors, and two bookshelves: one between the doors and the other along the wall to the left. Fresh air is introduced through a square opening in the center of the ceiling; one outflow vent is placed in the front part of the wall on the right, and another, near the back of the wall on the left just above the floor. In addition, the door on the left was open during the simulation.

In Figure 11.11, the flow is represented by the velocity vectors in a horizontal plane which are color-coded according to velocity magnitude. In addition, two streamlines are shown beginning on opposite sides of the inflow opening. The office furniture is shown transparently. One can see the principal flow regions as well as vorticity generation in front of the left door. One can also observe the complex winding paths taken by the flow.

In Figure 11.12 we see a velocity-magnitude isosurface colored according to the velocity component in the direction from left to right. Apart from showing the inflow of air, the coloring of the isosurface indicates the direction in which the flow moves in the room or is pushed away from the walls: blue denotes flow to the left; red, flow to the right.

Numerical investigations of room ventilation employing two different turbulence models can be found in [Sakamoto & Matsuo, 1980]. The numerical results there are also compared to physical experiments.

Flow through Technical Devices

In many applications the flow through a certain technical apparatus is of great interest. The flow patterns encountered here offer engineers important guidelines for the analysis and optimization of products.

Figure 11.13 shows a simple component of a technical device. The fluid enters

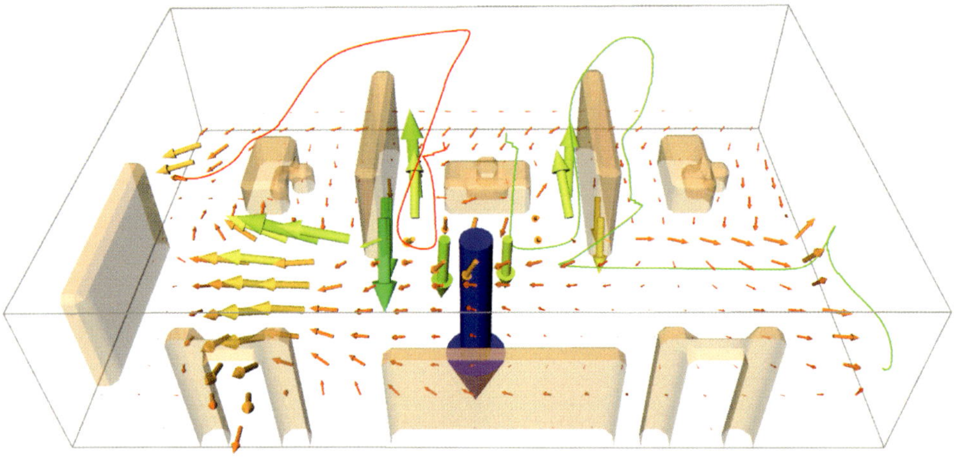

FIG. 11.11. *Room ventilation, velocity vectors in a horizontal plane color-coded according to velocity magnitude, and the course of two streamlines.*

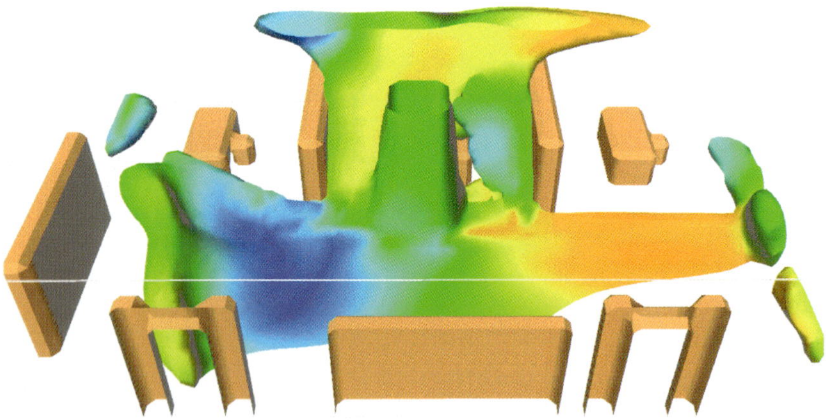

FIG. 11.12. *Room ventilation, isosurface of velocity magnitude color-coded according to the velocity component from left to right.*

on the left and exits on the top. The object shown here was generated using the *constructive solid geometry* (CSG) technique (in this case by intersecting an initial slab with a sphere and two cylinders). The results of a simulation at a Reynolds number of $Re = 1000$ and using a grid of $80 \times 40 \times 40$ cells are shown in Figure 11.13 by way of several visualization techniques in one picture. The object itself is indicated by its transparent shell, inside of which the flow field is visualized by a color-coding of the velocity magnitude in the central plane as well as colored arrows of different length. The course of the flow is indicated by three streamlines.

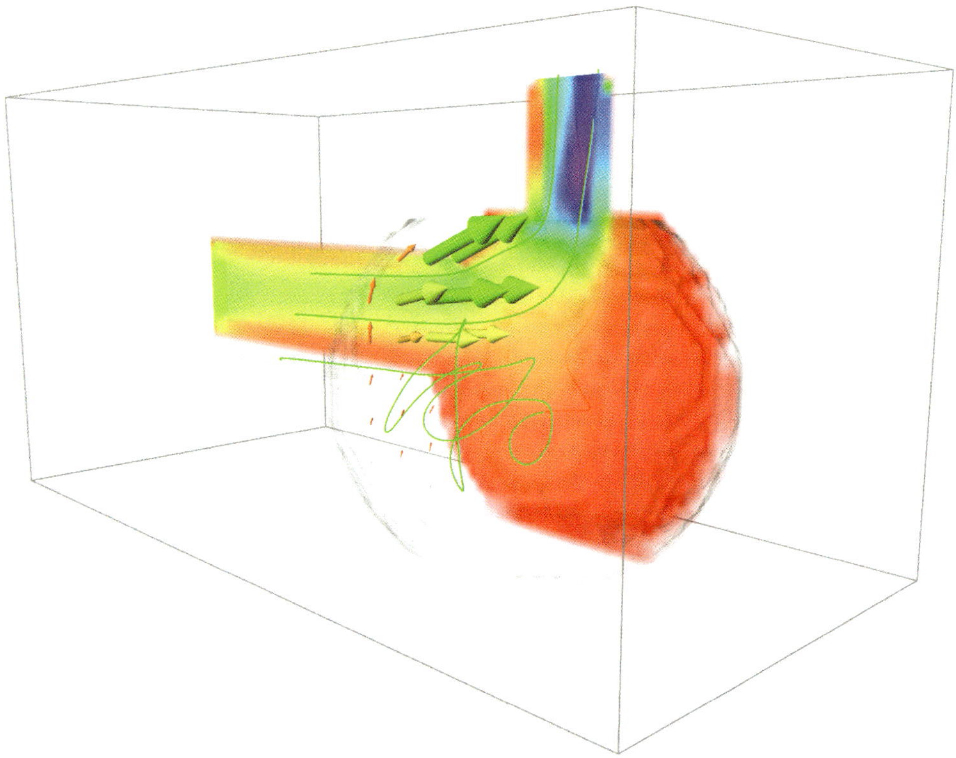

FIG. 11.13. *Flow through a component of a technical device.*

11.4.3 Examples of Free Boundary Value Problems

The Splash of a Liquid Drop

Taking up the falling drop example again (cf. Section 7.2, page 101), we now simulate this process in three dimensions. The viscosity of the fluid in the basin and in the drop are again the same ($Re = 12.5$). The calculation was done using the three-dimensional extension of the MAC technique described in Section 11.3, which was introduced for the two-dimensional case in Chapter 6. A resolution of $32 \times 32 \times 32$ cells was used including 64 particles per cell. Figure 11.14 no longer shows the interior of the fluid, only its surface.

To obtain a more realistic splashing process than a purely rotationally symmetric one, the initial data was given a small perturbation (in the form of a slight drop velocity in the x-direction). Figure 11.14 shows successive snapshots of the numerical experiment: in pictures 1 through 4, the droplet falls into the fluid-filled basin, impacts, and forms a trough. The first sign of asymmetry can be seen when the drop sloshes back up out of the basin (picture 5). This is followed by the formation of a swirling "fountain" (picture 6), which becomes unstable and falls over to the side (picture 7) before falling back into the basin (picture 8).

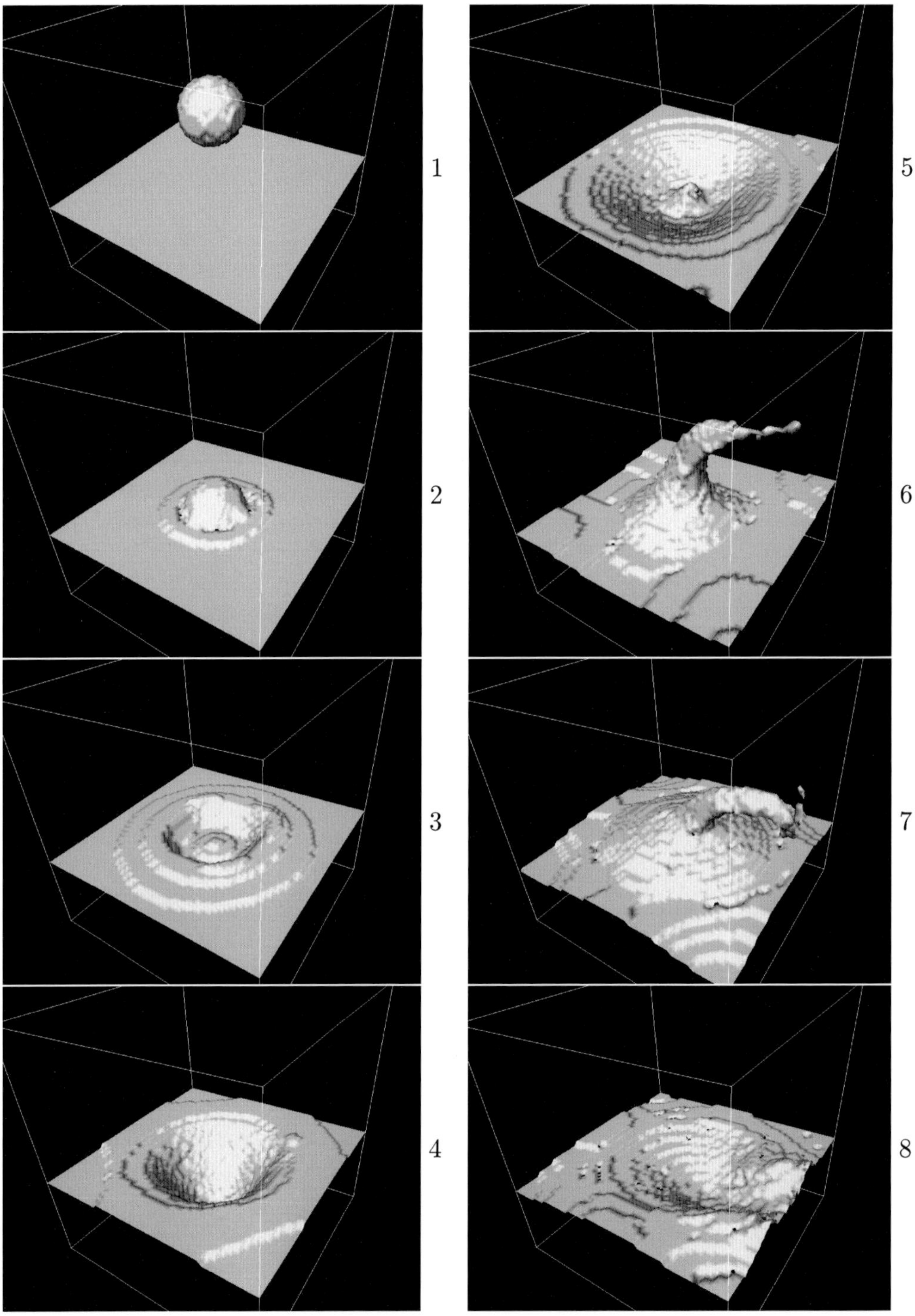

FIG. 11.14. *The splash of a liquid drop in three space dimensions.*

Sloshing Fluid

In a slight variation of the two-dimensional breaking dam simulation in Section 7.1, this simulation consists of placing a "fluid parcel" next to the left wall of a cube using a resolution of $32 \times 32 \times 32$ cells at a Reynolds number of $Re = 100$ and then suddenly removing the imaginary wall supporting the fluid. No-slip conditions are imposed along all sides of the cube. Hence, in contrast with the two-dimensional simulation in Section 7.1, the boundary on the right is no longer open. Figure 11.15 shows two snapshots of the sloshing fluid to briefly illuminate this process. Again, only the fluid surface is displayed. Due to the no-slip conditions at the wall, the fluid is seen to stick to the lateral walls in this illustration; i.e., it slides down the walls rather slowly.

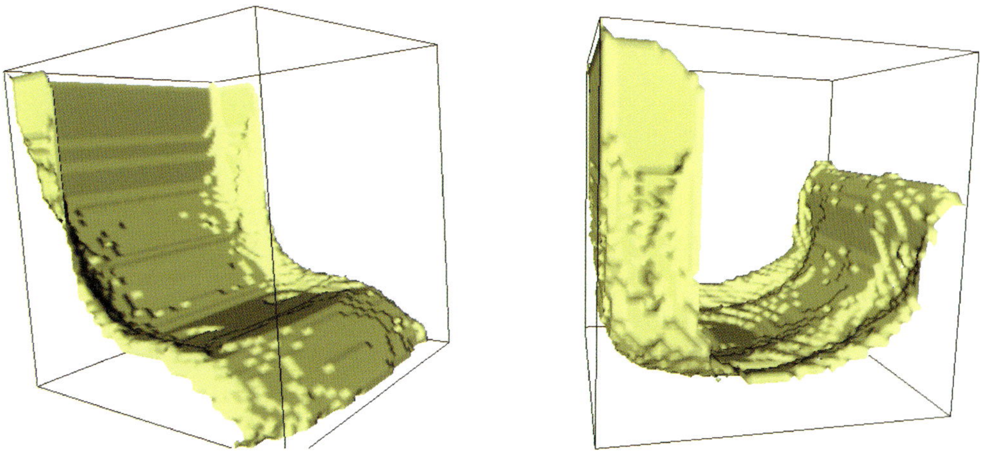

FIG. 11.15. *Sloshing fluid in three space dimensions.*

11.4.4 Examples of Temperature-Driven Flows

In Section 9.7.2, we have already treated natural convection in the two-dimensional case, where we studied in particular the development of convection cells (Rayleigh–Bénard cells). The two-dimensional simulation did not account for the third velocity component, which is justified only in two special cases: if the container is extended infinitely in one direction, making the effects of the lateral walls negligibly small, and for so-called Hele–Shaw flows, in which the distance between the horizontal walls is very small (cf. Section 9.7.2, page 140).

When the third spatial dimension is accounted for in the numerical simulation, one can observe in the simulation (as well as in physical experiments) that the convection flow inside a closed container is always three-dimensional. This is caused mainly by two mechanisms: on one hand, the interaction of the rotating fluid with the fixed sidewalls (with no-slip conditions) and, on the other hand, the temperature gradients, whose direction is orthogonal to the vertical walls (*thermal end effect*). Three-dimensional convection flows are treated extensively in the papers [Aziz & Hellums, 1967], [Grötzbach, 1983], [Kessler, 1987], and [Neumann, 1990].

In the following, we analyze natural convection in three different containers of varying depth. The physical parameters are those of numerical experiment 2 of Section 9.7.2 on page 142. In particular, we are interested in the number of convection cells as a function of the container geometry. The resolutions used are 64:8:16, 64:16:16, and 64:32:16, each of which is adapted to the corresponding container size.

Container with Length:Depth:Height = 8:1:2

In this flow, which closely approaches Hele–Shaw flow, six convection cells form, as in the two-dimensional experiment. This is seen in Figure 11.16 by observing the vector field in the central vertical plane. This picture also contains the isosurface of the reference temperature $T_\infty = 293$ K, which is color-coded with the vertical velocity component.

In Figure 11.17, the vector field is shown in a horizontal plane, and the color-coding of the temperature, in a vertical plane: warm fluid is shown in red; cold is shown in blue. Looking at both figures, one can clearly see how heat is transported with the flow inside the convection cells: it rises and descends with the flow.

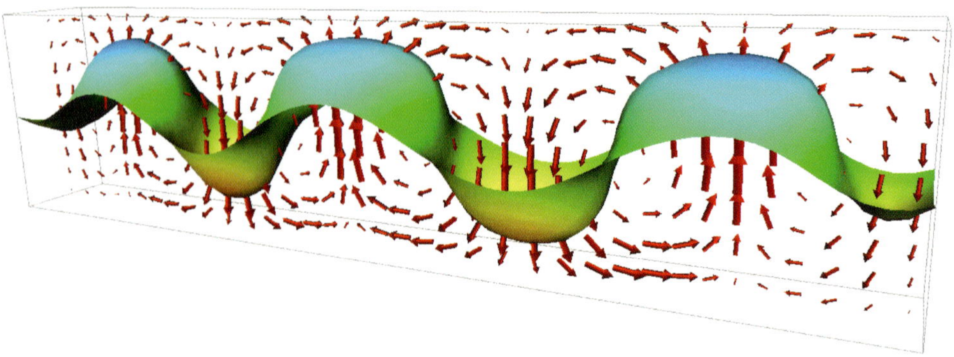

FIG. 11.16. *Container with length:depth:height* = 8:1:2, *vector field in a vertical central plane and isosurface of temperature* $T = T_\infty = 293$ *K color-coded according to the vertical velocity component.*

Container with Length:Depth:Height = 8:2:2

The number of convection cells reduces from six to four in this case due to the growing influence of the left and right vertical walls. This is illustrated in Fig-

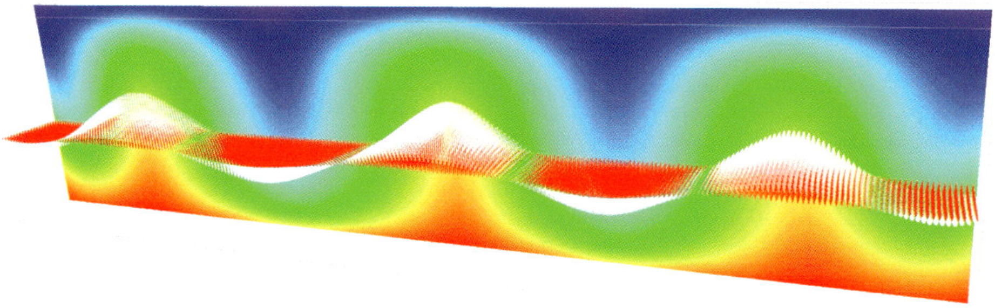

FIG. 11.17. *Container with length:depth:height* = 8:1:2, *vector field in a horizontal vertical plane and color-coding of temperature in a vertical plane.*

ure 11.18, in which the vector field is again shown in a vertical plane accompanied by the color coding of the upward velocity component in a plane parallel to the first. The combination of these two representations gives an impression of the position and structure of the convection cells.

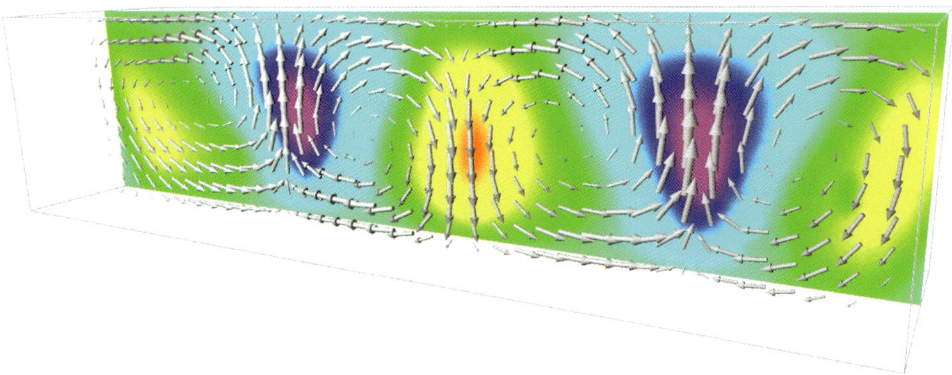

FIG. 11.18. *Length:depth:height* = 8:2:2, *vector field and color-coding of upward velocity component in two parallel planes.*

Container with Length:Depth:Height = 8:4:2

The container in this case has the largest lateral walls of the three, and their effect is seen to be the strongest here as well: only three cells develop (cf. [Kessler, 1987]). This can be seen in Figure 11.19, in which the vector field is again shown in a vertical plane, and the color-coded temperature, in a horizontal plane. The color-coding also clearly shows the cold fluid being transported downward and the warm fluid being transported upward in the convection cells.

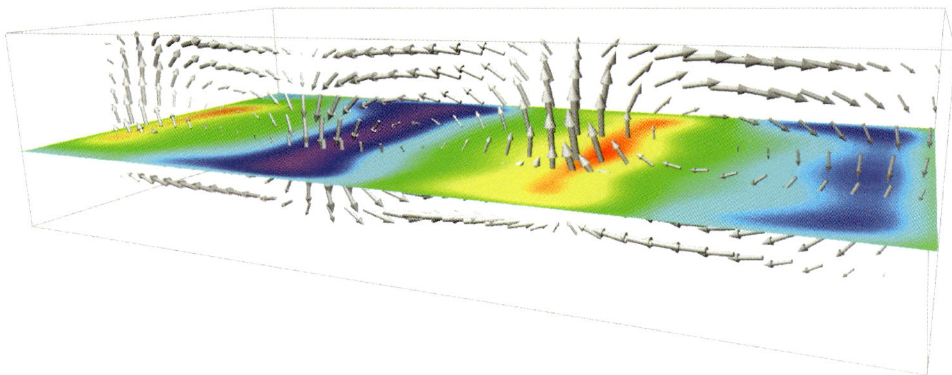

FIG. 11.19. *Container with length:depth:height* = 8:4:2, *vector field in a vertical plane and color-coded temperature in a horizontal plane.*

Figure 11.20 also reveals a new phenomenon distinguishing the two- and three-dimensional experiments in an essential way. The streamlines now no longer remain confined to a single convection cell; rather, there is an exchange of fluid among the cells—in contrast with the two-dimensional experiment (cf. configuration 2) of Section 9.7.2 on page 142.

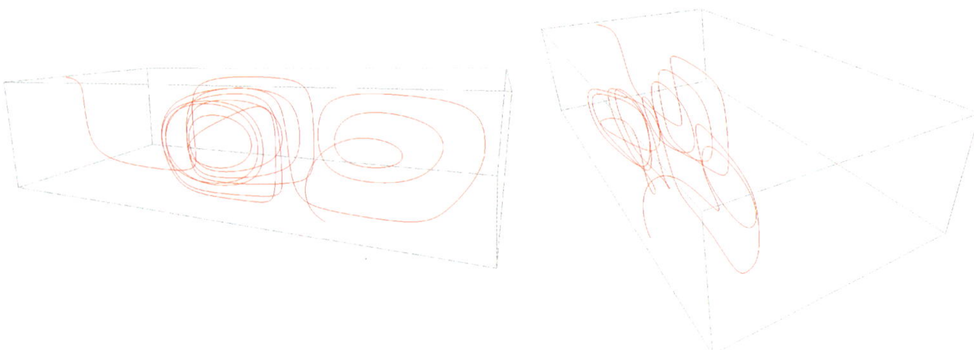

FIG. 11.20. *Container with length:depth:height* = 8:4:2, *streamline winding through all three convection cells.*

Finally, the area between two Rayleigh–Bénard cells, in which the fluid exchange occurs from one cell to the next, is visualized in Figure 11.21 by displaying the velocity vector field in a vertical plane which separates the two cells.

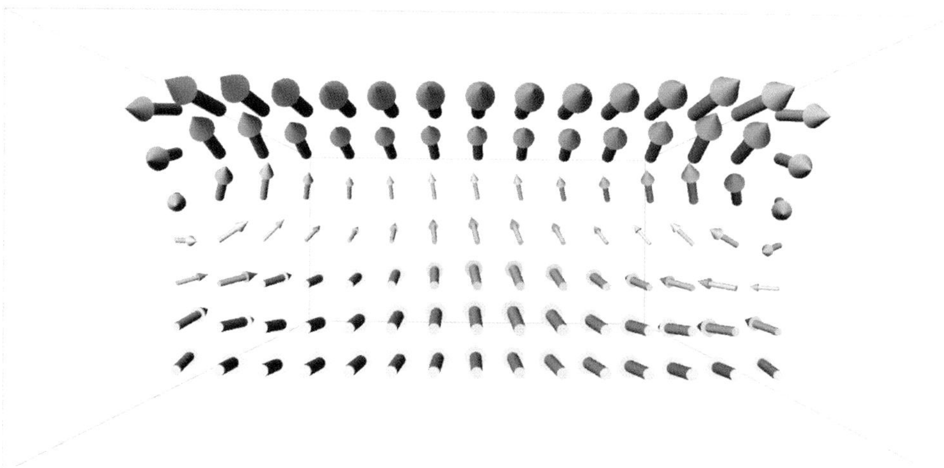

FIG. 11.21. *Container with length:depth:height* = 8:4:2, *velocity vector field in a vertical plane separating two convection cells.*

Chapter 12

Concluding Remarks

In this book we have exhibited the basic steps of numerical simulation based on a simple algorithm for the simulation of unsteady, laminar, viscous, incompressible flow. Our objective was to enable readers to understand and implement a code for the numerical solution of the Navier–Stokes equations. Beyond this, we have also described the use of modern tools such as visualization to aid in interpretation of the computed data and parallelization on often easily accessible networks of workstations. Additional chapters on energy transport, turbulence, and three-dimensional simulations supply the necessary background for approaching problems of practical relevance.

We are aware of the existence of substantially more efficient numerical methods, such as the use of semi-implicit or implicit time discretizations, higher-order methods for the spatial discretization, and the application of body-fitted grids using generalized coordinates, block-structured, or even completely unstructured grids.[87] Moreover, the acceleration of the calculations by multigrid methods, the use of adaptive refinement techniques controlled by error estimators, and improved parallelization strategies including load balancing are all current areas of research.

Not least of all, we point out that the validation of simulation results is an important task requiring intensive collaboration with the scientists or engineers conducting the physical experiments.

The fast-paced development of the computer hardware market will further increase the opportunities for numerical simulation, and hence also the requirements placed on scientific computing. We are thus convinced that application areas for numerical simulation will become increasingly varied in the future and that numerical simulation will grow to become an essential tool in all areas of science and engineering.

We hope that readers, having worked their way through the book, will have an easier time finding their way around other application areas of numerical simulation besides fluid dynamics and will be able to successfully apply the scientific computing techniques learned in this book to these.

[87]In addition, there exist fundamentally different methods of discretization such as finite volume and finite element methods. For these we refer to the relevant literature.

Appendix A

Guidelines for Parallelization Using PVM

Before calling PVM, certain file paths containing the PVM daemon as well as the executable PVM programs must be correctly set. Readers should consult their local system administrators to find out what these should be on their systems.

Once this is done, PVM may be started using the command `pvm`. This command starts up the PVM console, a command interpreter from which commands such as adding a new host to the virtual machine, spawning processes, or obtaining information on the current configuration may be submitted. The command `help` lists a short summary of available commands.

Another option is to start the daemon directly on all hosts listed in a file `hostfile` using the command `pvmd hostfile &` . Each line in `hostfile` should have the form

hostname `ep=`*path1* `dx=`*path2*`/pvmd`,

in which *path1* is the path to the executable program on the corresponding host and *path2* is the path to the PVM daemon.

When the daemons are no longer needed, they should be terminated by submitting the `halt` command from the PVM console.

Our parallel flow computations require the following PVM library functions:

- `int tid = pvm_mytid()` returns an integer, the so-called *task identifier* (`tid`), which uniquely identifies the calling process in the PVM system. Negative return values indicate an error. It is common practice to call this PVM function before any other.
- `int tid = pvm_parent()` returns the `tid` of the process which spawned the calling process. If the calling process was not spawned by another process, e.g., if it is the master process, the value `PvmNoParent`, a macro defined in the header file `pvm3.h`, is returned.
- `int numt = pvm_spawn(char *task,char **argv,int flag,char *where, int ntask,int *tids)` spawns `ntask` copies of the executable `task` on the host or architecture specified in `where`. The parameter `flag` indicates whether an arbitrary host may be selected (`flag=0,where=" "`), whether a specific host is specified in `where` (`flag=1`), or whether a specific architecture is specified in `where` (`flag=2`). In particular, `flag` is set to 1 if a

specific available connection topology (e.g., tree-like) is to be used. The task identifiers of the `ntask`-spawned processes are returned in `tids`, and `argv` contains the command line arguments to be supplied in the call of `task`. The return value is `numt`, the number of processes actually spawned.

- `int bufid = pvm_initsend(int encoding)` prepares the sending of a new message. The parameter `encoding` specifies how the data to be sent is encoded. To send a message between hosts of different architectures, `encoding` must have the value `PvmDataDefault`; for messages to hosts of the same architecture the value `PvmDataRaw` can be used. Both values are macros defined in `pvm3.h`. A pointer to the buffer to be used for sending the message is returned in `bufid`.

- `int info = pvm_pk`*type*`(`*type* `*p,int nitem,int stride)` writes the data to be sent into the buffer prepared by `pvm_initsend`. Here *type* denotes the data type of the data to be sent. Possible types are, e.g., `int`, `float`, or `double`. The number of data elements is given by `nitem`, `stride` denotes the stride to use when packing, and `p` is a pointer to the first data element to be sent. The array `p` must contain at least `nitem` · `stride` elements. A negative return value `info` signals the occurrence of an error.

- `int info = pvm_upk`*type*`(`*type* `*p,int nitem,int stride)` copies the data from the message buffer to the array `p`. The parameters have the same meanings as in `pvm_pk`*type*.

- `int info = pvm_mcast(int *tids,int ntask,int msgtag)` sends the data contained in the message buffer to `ntask` processes specified in `tids`. The nonnegative message tag `msgtag` allows distinguishing between different messages.

- `int info = pvm_send(int tid,int msgtag)` sends the data contained in the message buffer to the process specified in `tid`. The parameter `msgtag` has the same function as in `pvm_mcast`.

- `int bufid = pvm_recv(int tid,int msgtag)` waits until a message sent by process `tid` carrying the message tag `msgtag` arrives and places the data into the message buffer `bufid`.

- `int info = pvm_exit()` informs the daemon that the calling process is leaving PVM.

Several calls to `pvm_pk`*type* may follow `pvm_initsend` in order to send data from several arrays in one communication step. However, it is recommended that data of the same type first be placed in an intermediate buffer, followed by one call of `pvm_pk`*type* for this particular type.

The source code of a simple PVM program taken from the PVM Reference Manual [Geist et al., 1995] is given below, in which, following an initialization part, a token is passed around once in a circle of NPROC hosts.

```
#define NPROC 4
#include "pvm3.h"
main()
{
    int mytid,tids[NPROC],me,i;

    mytid = pvm_mytid();          /* enroll in pvm */
    tids[0] = pvm_parent();       /* find out if I am parent or child */
    if( tids[0] < 0 ){            /* then I am the parent */
       tids[0] = mytid;
       me = 0;                    /* start up copies of myself */
       pvm_spawn("a.out", (char**)0, 0, "", NPROC-1, &tids[1]);
       pvm_initsend( PvmDataDefault );      /* multicast tids array */
       pvm_pkint(tids, NPROC, 1);           /* to children */
       pvm_mcast(&tids[1], NPROC-1, 0);}
    else{                         /* I am a child */
       pvm_recv(tids[0], 0);      /* receive tids array */
       pvm_upkint(tids, NPROC, 1);
       for( i=1; i<NPROC ; i++ )
          if( mytid == tids[i] ){ me = i; break; }}
/*-----------------------------------------------------------
 * all NPROC tasks are equal now
 * and can address each other by tids[0] thru tids[NPROC-1]
 * for each process me => process number [0-(NPROC-1)]
 *-----------------------------------------------------------*/
     dowork( me, tids, NPROC );
     pvm_exit();                  /* program finished exit pvm */
     exit(1);
}
                                /* dowork passes a token around a ring */
dowork(int  me,int*  tids,int nproc )
{
     int token, dest, count  = 1, stride = 1, msgtag = 4;
     if( me == 0 ){
        token = tids[0];
        pvm_initsend( PvmDataDefault );
        pvm_pkint( &token, count, stride );
        pvm_send( tids[me+1], msgtag );
        pvm_recv( tids[nproc-1], msgtag );
        printf("token ring done\n"); }
     else {
        pvm_recv( tids[me-1], msgtag );
        pvm_upkint( &token, count, stride );
        pvm_initsend( PvmDataDefault );
        pvm_pkint( &token, count, stride );
        dest = (me == nproc-1)? tids[0] : tids[me+1] ;
        pvm_send( dest, msgtag ); }
}
```

Appendix B

Physical Properties of Fluids

Values for Water at Atmospheric Pressure

Temperature T [°C]	Density $\varrho\left[\frac{kg}{m^3}\right]$	Dynamic viscosity $\mu\left[\frac{kg}{m\,s}\right]$	Kinematic viscosity $\nu\left[\frac{m^2}{s}\right]$	Thermal conductivity $\kappa\left[\frac{W}{m\,K}\right]$	Thermal diffusivity $\alpha\left[\frac{m^2}{s}\right]$	Coefficient of thermal expansion $\beta\left[\frac{1}{K}\right]$	Prandtl number $Pr=\frac{\nu}{\alpha}$
0	999.9	$17.87\cdot10^{-4}$	$17.87\cdot10^{-7}$	0.56	$1.33\cdot10^{-7}$	$-0.6\cdot10^{-4}$	13.44
5	1000	$15.14\cdot10^{-4}$	$15.14\cdot10^{-7}$	0.57	$1.36\cdot10^{-7}$	$+0.1\cdot10^{-4}$	11.13
10	999.7	$13.04\cdot10^{-4}$	$13.04\cdot10^{-7}$	0.58	$1.38\cdot10^{-7}$	$0.9\cdot10^{-4}$	9.45
15	999.1	$11.37\cdot10^{-4}$	$11.38\cdot10^{-7}$	0.59	$1.40\cdot10^{-7}$	$1.5\cdot10^{-4}$	8.13
20	998.2	$10.02\cdot10^{-4}$	$10.04\cdot10^{-7}$	0.59	$1.42\cdot10^{-7}$	$2.1\cdot10^{-4}$	7.07
25	997.1	$8.91\cdot10^{-4}$	$8.94\cdot10^{-7}$	0.60	$1.44\cdot10^{-7}$	$2.6\cdot10^{-4}$	6.21
30	995.7	$7.98\cdot10^{-4}$	$8.02\cdot10^{-7}$	0.61	$1.46\cdot10^{-7}$	$3.0\cdot10^{-4}$	5.49
35	994.1	$7.20\cdot10^{-4}$	$7.25\cdot10^{-7}$	0.62	$1.49\cdot10^{-7}$	$3.4\cdot10^{-4}$	4.87
40	992.3	$6.54\cdot10^{-4}$	$6.59\cdot10^{-7}$	0.63	$1.52\cdot10^{-7}$	$3.8\cdot10^{-4}$	4.34
50	988.1	$5.48\cdot10^{-4}$	$5.54\cdot10^{-7}$	0.64	$1.55\cdot10^{-7}$	$4.5\cdot10^{-4}$	3.57
60	983.2	$4.67\cdot10^{-4}$	$4.75\cdot10^{-7}$	0.65	$1.58\cdot10^{-7}$	$5.1\cdot10^{-4}$	3.01
70	977.8	$4.05\cdot10^{-4}$	$4.14\cdot10^{-7}$	0.66	$1.61\cdot10^{-7}$	$5.7\cdot10^{-4}$	2.57
80	971.8	$3.55\cdot10^{-4}$	$3.66\cdot10^{-7}$	0.67	$1.64\cdot10^{-7}$	$6.2\cdot10^{-4}$	2.23
90	965.3	$3.16\cdot10^{-4}$	$3.27\cdot10^{-7}$	0.67	$1.65\cdot10^{-7}$	$6.7\cdot10^{-4}$	1.98
100	958.4	$2.83\cdot10^{-4}$	$2.95\cdot10^{-7}$	0.68	$1.66\cdot10^{-7}$	$7.1\cdot10^{-4}$	1.78

Values for Air at Atmospheric Pressure

Temperature T [°C]	Density $\varrho\left[\frac{kg}{m^3}\right]$	Dynamic viscosity $\mu\left[\frac{kg}{m\,s}\right]$	Kinematic viscosity $\nu\left[\frac{m^2}{s}\right]$	Thermal conductivity $\kappa\left[\frac{W}{m\,K}\right]$	Thermal diffusivity $\alpha\left[\frac{m^2}{s}\right]$	Coefficient of thermal expansion $\beta\left[\frac{1}{K}\right]$	Prandtl number $Pr=\frac{\nu}{\alpha}$
-180	3.720	$0.65\cdot10^{-5}$	$0.175\cdot10^{-5}$	$0.76\cdot10^{-4}$	$0.19\cdot10^{-5}$	$1.08\cdot10^{-2}$	0.92
-100	2.040	$1.16\cdot10^{-5}$	$0.57\cdot10^{-5}$	$1.6\cdot10^{-4}$	$0.76\cdot10^{-5}$	$5.74\cdot10^{-3}$	0.75
-50	1.582	$1.45\cdot10^{-5}$	$0.92\cdot10^{-5}$	-	-	-	-
0	1.293	$1.71\cdot10^{-5}$	$1.32\cdot10^{-5}$	$2.4\cdot10^{-4}$	$1.84\cdot10^{-5}$	$3.66\cdot10^{-3}$	0.72
10	1.247	$1.76\cdot10^{-5}$	$1.41\cdot10^{-5}$	$2.5\cdot10^{-4}$	$1.96\cdot10^{-5}$	$3.52\cdot10^{-3}$	0.72
20	1.205	$1.81\cdot10^{-5}$	$1.50\cdot10^{-5}$	$2.5\cdot10^{-4}$	$2.08\cdot10^{-5}$	$3.40\cdot10^{-3}$	0.72
30	1.165	$1.86\cdot10^{-5}$	$1.60\cdot10^{-5}$	-	-	-	-
60	1.060	$2.00\cdot10^{-5}$	$1.88\cdot10^{-5}$	-	-	-	-
100	0.946	$2.18\cdot10^{-5}$	$2.30\cdot10^{-5}$	$3.2\cdot10^{-4}$	$3.28\cdot10^{-5}$	$2.68\cdot10^{-3}$	0.7
200	0.746	$2.58\cdot10^{-5}$	$3.46\cdot10^{-5}$	-	-	-	-
300	0.616	$2.95\cdot10^{-5}$	$4.81\cdot10^{-5}$	-	-	-	-
500	0.456	$3.58\cdot10^{-5}$	$7.85\cdot10^{-5}$	-	-	-	-
1000	0.277	$4.82\cdot10^{-5}$	$17.4\cdot10^{-5}$	$7.6\cdot10^{-4}$	$27.1\cdot10^{-5}$	$7.83\cdot10^{-4}$	0.64

Bibliography

[Almasi & Gottlieb, 1994] Almasi, G. & Gottlieb, A. (1994). *Highly Parallel Computing.* Redwood City: Benjamin/Cummings, 2nd edition.

[Amsden & Harlow, 1970a] Amsden, A. & Harlow, F. (1970a). A simplified MAC technique for incompressible fluid flow calculations. *J. Comput. Phys.*, 6, 322–325.

[Amsden & Harlow, 1970b] Amsden, A. & Harlow, F. (1970b). *The SMAC Method: A Numerical Technique for Calculating Incompressible Fluid Flow.* Technical report LA-4370, Los Alamos, NM: Los Alamos National Laboratory.

[Anderson et al., 1984] Anderson, D., Tannehill, J., & Pletcher, R. (1984). *Computational Fluid Mechanics and Thermal Science.* Washington: Hemisphere Publishers.

[Armaly et al., 1983] Armaly, B., Durst, F., Pereira, J., & Schönung, B. (1983). Experimental and theoretical investigation of backward-facing step flow. *J. Fluid. Mech.*, 127, 473–496.

[Aziz & Hellums, 1967] Aziz, K. & Hellums, J. (1967). Numerical solution of the three-dimensional equations of motion for laminar natural convection. *Phys. Fluids*, 10, 314–324.

[Babuška et al., 1986] Babuška, I., Zienkiewicz, O., Gago, L., & Olivera, E., Eds. (1986). *Accuracy Estimates and Adaptive Refinement in Finite Element Computations.* New York: John Wiley.

[Bader, 1995] Bader, M. (1995). *Berechnung turbulenter Strömungen mit dem k-ε-Modell.* Fortgeschrittenenpraktikum, Institut für Informatik, TU München.

[Bank, 1994] Bank, R. (1994). *PLTMG: A Software Package for Solving Elliptic Partial Differential Equations, Users Guide* 7.0. Philadelphia: SIAM.

[Barrett et al., 1993] Barrett, R., Berry, M., Chan, T., Demmel, J., Donato, J., Dongarra, J., Eijckhout, V., Pozo, R., Romine, C., & van der Vorst, H. (1993). *Templates for the Solution of Linear Systems.* Philadelphia: SIAM.

[Bastian, 1993] Bastian, P. (1993). *Parallel Adaptive Multigrid Methods.* Preprint 93–60, IWR Heidelberg.

[Bauer & Wössner, 1982] Bauer, F. & Wössner, H. (1982). *Algorithmic Language and Program Development.* Berlin: Springer-Verlag.

[Bejan, 1984] Bejan, A. (1984). *Convection Heat Transfer.* New York: Wiley-Interscience.

[Berger & Gastiaux, 1988] Berger, M. & Gastiaux, B. (1988). *Differential Geometry: Manifolds, Curves and Surfaces.* New York: Springer-Verlag.

[Bernsdorf, 1994] Bernsdorf, J. (1994). *Computersimulation von Strömungen in komplexen Geometrien mittels Lattice-Boltzmann-Automaten.* Diplomarbeit, Universität Oldenburg.

[Blum, 1993] Blum, J. (1993). Oben hui und unten pfui. *Geo*, 8, 10–34.

[Bode & Händler, 1983] Bode, A. & Händler, W. (1983). *Rechnerarchitektur* II. Berlin: Springer-Verlag.

[Boussinesq, 1903] Boussinesq, J. (1903). *Theorie Analytique de la Chaleur.* Paris: Gauthier-Villars, 2.

[Brandt, 1984] Brandt, A. (1984). *Multigrid Techniques:* 1984 *Guide with Applications to Fluid Dynamics.* Technical report, GMD-Studie 85, St. Augustin.

[Bräunl, 1993] Bräunl, T. (1993). *Parallel Programming.* New York: Prentice Hall.

[Braza et al., 1986] Braza, M., Chassing, P., & Minh, H. H. (1986). Numerical study and physical analysis of the pressure and velocity fields in the near wake of a circular cylinder. *J. Fluid Mech.*, 165, 79–130.

[Brenner & Scott, 1994] Brenner, S. & Scott, L. R. (1994). *The Mathematical Theory of Finite Element Methods.* New York: Springer-Verlag.

[Brezzi, 1974] Brezzi, F. (1974). On the existence, uniqueness and approximation of saddle-point problems arising from Lagrangian multipliers. *RAIRO Anal. Numér.*, 8, R-2, 129–151.

[Browne, 1978] Browne, L. (1978). The marker and cell technique. In J. Noye (Ed.), *Numerical Simulation of Fluid Motion.* Amsterdam: North-Holland, 223–247.

[Bungartz & Schulte, 1995] Bungartz, H.-J. & Schulte, S. (1995). *Coupled Problems in Microsystem Technology.* SFB-Bericht 342/07/95 A, Institut für Informatik, TU München.

[Cantwell, 1981] Cantwell, B. (1981). Organized motion in turbulent flow. *Ann. Rev. Fluid Mech.*, 13, 457–515.

[Canuto et al., 1988] Canuto, C., Hussaini, M., Quarteroni, A., & Zhang, T. (1988). *Spectral Methods in Fluid Dynamics.* Berlin: Springer-Verlag.

[Chan & Mathew, 1994] Chan, T. & Mathew, T. (1994). Domain decomposition algorithms. *Acta Numerica*, 61–143.

[Chang et al., 1991] Chang, K., Chen, C., & Uang, C. (1991). A hybrid k-ε model of recirculating flow. *Internat. J. Numer. Methods Fluids*, 12, 369–382.

[Chase, Jr. et al., 1985] Chase, Jr., M., Davies, C., Downey, J., Frurip, D., McDonald, R., & Syverud, A. (1985). Jannaf thermochemical tables, 3rd edition. *J. Phys. Chem. Ref. Data*, 14, 1–1856.

[Chen et al., 1991] Chen, S., Johnson, D., & Raad, P. (1991). The surface marker method. In L. Wrobel & C. Brebbia (Eds.), *Computational Modelling of Free and Moving Boundary Problems, Vol.* 1 *Fluid Flow*. Berlin: de Gruyter, 223–235.

[Chorin, 1968] Chorin, A. (1968). Numerical solution of the Navier-Stokes-equations. *Math. Comp.*, 22, 745–762.

[Chorin & Marsden, 1979] Chorin, A. & Marsden, J. (1979). *A Mathematical Introduction to Fluid Mechanics*. New York: Springer-Verlag.

[Ciarlet, 1978] Ciarlet, P. (1978). *The Finite Element Method for Elliptic Problems*. Amsterdam: North-Holland.

[Crank & Nicholson, 1947] Crank, J. & Nicholson, P. (1947). A practical method for the numerical evaluation of solutions of partial differential equations of the heat-conduction type. *Proc. Cambridge Philos. Soc.*, 43, 50–67.

[Daly & Pracht, 1968] Daly, J. & Pracht, W. (1968). Numerical study of density-current surges. *Phys. Fluids*, 11, 15–30.

[Dennis, 1985] Dennis, S. (1985). Compact explicit finite-difference approximation to Navier-Stokes equations. In Soubbaramayer & J. Boujot (Eds.), 9*th Internat. Conf. Numer. Methods in Fluid Dynamics, Lecture Notes in Physics, Vol.* 128. Berlin, Heidelberg: Springer-Verlag, 23–36.

[Dennis & Chang, 1970] Dennis, S. & Chang, G. (1970). Numerical solutions for steady flow past a circular cylinder at Reynolds numbers up to 100. *J. Fluid Mech.*, 42, 471–489.

[Dennis et al., 1979] Dennis, S., Ingham, D., & Cook, R. (1979). Finite-difference methods for calculating steady incompressible flows in three dimensions. *J. Comput. Phys.*, 33, 325–339.

[Douglas & Hornung, 1993] Douglas, J. & Hornung, U., Eds. (1993). *Flow in Porous Media*. Basel: Birkhäuser.

[Driver & Seegmüller, 1985] Driver, D. & Seegmüller, H. (1985). Features of a reattaching turbulent shear layer in divergent channel flows. *AIAA Journal*, 23, 163–171.

[Durst, 1991] Durst, B. (1991). *Numerische Simulation einer periodischen Nischenströmung unter Verwendung dünner Gitter*. Diplomarbeit, Institut für Informatik, TU München.

[Durst, 1994] Durst, F. (1994). High performance scientific computing and its application in solving engineering problems. In M. Griebel & C. Zenger (Eds.), *Numerical Simulation in Science and Engineering*. Braunschweig: Vieweg, 39–51.

[Eaton, 1987] Eaton, B. (1987). Analysis of laminar vortex shedding behind a circular cylinder by computer-aided flow visualization. *J. Fluid Mech.*, 180, 117–145.

[Eggels et al., 1994] Eggels, J., Unger, F., Weiss, M., Westerweel, J., Adrian, R., & Friedrich, R. (1994). Fully developed turbulent pipe flow: A comparison between direct numerical simulation and experiments. *J. Fluid Mech.*, 268, 175–209.

[Ern & Giovangigli, 1994] Ern, A. & Giovangigli, V. (1994). *Multicomponent Transport Algorithms*. Heidelberg: Springer-Verlag.

[Farge, 1992] Farge, M. (1992). Wavelet transforms and their applications to turbulence. *Ann. Rev. Fluid Mech.*, 24, 395–457.

[Fischer & Troger, 1994] Fischer, H. & Troger, C. (1994). Computational fluid dynamics with FIRE on massive parallel computers. In M. Griebel & C. Zenger (Eds.), *Numerical Simulation in Science and Engineering*. Braunschweig: Vieweg, 52–66.

[Fletcher, 1991] Fletcher, C. (1991). *Computational Techniques for Fluid Dynamics* I & II. Berlin: Springer-Verlag, 2nd edition.

[Foley & van Dam, 1984] Foley, J. & van Dam, A. (1984). *Fundamentals of Interactive Computer Graphics*. London, Amsterdam: Addison-Wesley.

[Gartling, 1990] Gartling, D. (1990). A test problem for outflow boundary conditions—flow over a backward-facing step. *Internat. J. Numer. Methods Fluids*, 11, 511–537.

[Gatski & Grosch, 1985] Gatski, T. & Grosch, C. (1985). Numerical study of the two- and three-dimensional unsteady Navier–Stokes equations in velocity-vorticity variables using compact difference schemes. In Soubbaramayer & J. Boujot (Eds.), *9th Internat. Conf. Numer. Methods in Fluid Dynamics, Lecture Notes in Physics, Vol.* 128. Berlin, Heidelberg: Springer-Verlag, 235–239.

[Gatski et al., 1982] Gatski, T., Grosch, C., & Rose, M. (1982). A numerical study of the two-dimensional Navier–Stokes equations in vorticity-velocity variables. *J. Comput. Phys.*, 48, 1–22.

[Gatski et al., 1989] Gatski, T., Grosch, C., & Rose, M. (1989). The numerical solution of the Navier–Stokes equations for 3-dimensional unsteady, incompressible flows by compact schemes. *J. Comput. Phys.*, 82, 298–329.

[Geist et al., 1994] Geist, A., Dongarra, J., Jiang, W., Manchek, R., & Sunderam, V. (1994). *PVM: Parallel Virtual Machine, A Users' Guide and Tutorial for Networked Parallel Computing*. Cambridge, MA: MIT Press.

[Geist et al., 1995] Geist, A., Dongarra, J., Jiang, W., Manchek, R., & Sunderam, V. (1995). *PVM3 User's Guide and Reference Manual*. Available from `ftp netlib2.cs.utk.edu pvm3/ug.ps`.

[Gentry et al., 1966] Gentry, R., Martin, R., & Daly, B. (1966). An Eulerian differencing method for unsteady compressible flow problems. *J. Comput. Phys.*, 1, 87–118.

[Germano, 1992] Germano, M. (1992). Turbulence: The filtering approach. *J. Fluid Mech.*, 238, 325–336.

[Ghia et al., 1982] Ghia, U., Ghia, K., & Shin, C. (1982). High-Re solutions for incompressible flow using the Navier–Stokes equations and a multigrid method. *J. Comput. Phys.*, 48, 387–411.

[Gibson & Launder, 1978] Gibson, M. & Launder, B. (1978). Ground effects on pressure fluctuations in the atmospheric boundary layer. *J. Fluid Mech.*, 86, 491–511.

[Golub & Ortega, 1992] Golub, G. & Ortega, J. (1992). *Scientific Computing and Differential Equations.* Boston: Academic Press.

[Gray & Giorgini, 1976] Gray, D. & Giorgini, A. (1976). On the validity of the Boussinesq approximation for liquids and gases. *J. Heat Mass Transfer*, 19, 545–551.

[Griebel & Huber, 1994] Griebel, M. & Huber, W. (1994). *Turbulence Simulation on Sparse Grids Using the Combination Technique.* SFB-Bericht 342/19/94 A, Institut für Informatik, TU München.

[Griebel et al., 1997] Griebel, M., Merz, W., & Neunhoeffer, T. (1997). *Mathematical Modeling and Numerical Simulation of Freezing Processes of a Supercooled Melt under Consideration of Density Changes*, Universität Bonn.

[Gropp et al., 1994] Gropp, W., Lusk, E., & Skjellum, A. (1994). *Using MPI, Portable Parallel Programming with the Message-Passing Interface.* Cambridge, MA: MIT Press.

[Grötzbach, 1983] Grötzbach, G. (1983). Spatial resolution requirements for direct numerical simulation of the Rayleigh–Bénard convection. *J. Comput. Phys.*, 49, 241–264.

[Hackbusch, 1985] Hackbusch, W. (1985). *Multi-Grid Methods and Applications.* Berlin: Springer-Verlag.

[Hackbusch, 1992] Hackbusch, W. (1992). *Elliptic Differential Equations, Theory and Numerical Treatment.* Berlin: Springer-Verlag.

[Hackbusch, 1994] Hackbusch, W. (1994). *Iterative Solution of Large Sparse Linear Systems of Equations.* Berlin: Springer-Verlag.

[Hammache & Gharib, 1991] Hammache, M. & Gharib, M. (1991). An experimental study of the parallel and oblique vortex shedding from circular cylinders. *J. Fluid Mech.*, 232, 567–590.

[Hanjalic & Launder, 1976] Hanjalic, K. & Launder, B. (1976). Contribution towards a Reynolds-stress closure for low-Reynolds number turbulence. *J. Fluid. Mech.*, 74, 593–610.

[Harlow & Fromm, 1965] Harlow, F. & Fromm, J. (1965). Computer experiments in fluid dynamics. *Scientific American*, 212, 104–110.

[Harlow & Shannon, 1967a] Harlow, F. & Shannon, J. (1967a). Distortion of a splashing liquid drop. *Science*, 157, 547–550.

[Harlow & Shannon, 1967b] Harlow, F. & Shannon, J. (1967b). The splash of a liquid drop. *J. Appl. Phys.*, 38, 3855–3866.

[Harlow & Welch, 1965] Harlow, F. & Welch, J. (1965). Numerical calculation of time-dependent viscous incompressible flow of fluid with free surface. *Phys. Fluids*, 8, 2182–2189.

[Harrington, 1983] Harrington, S. (1983). *Computer Graphics – a Programming Approach.* New York: McGraw-Hill.

[Hermann, 1980] Hermann, A. (1980). *Weltreich der Physik.* Esslingen: Bechtle.

[Hinze, 1975] Hinze, J. (1975). *Turbulence.* New York: McGraw-Hill.

[Hirasaki & Hellums, 1970] Hirasaki, G. & Hellums, J. (1970). Boundary conditions on the vector and scalar potentials in viscous three-dimensional hydrodynamics. *Quart. Appl. Math.*, 28, 293.

[Hirt & Cook, 1972] Hirt, C. & Cook, J. (1972). Calculating three-dimensional flows around structures and over rough terrain. *J. Comput. Phys.*, 10, 324–340.

[Hirt & Nichols, 1981] Hirt, C. & Nichols, B. (1981). Volume of fluid (VOF) method for the dynamics of free boundaries. *J. Comput. Phys.*, 39, 201–225.

[Hirt et al., 1975] Hirt, C., Nichols, B., & Romero, N. (1975). *SOLA – A Numerical Solution Algorithm for Transient Fluid Flows.* Technical report LA-5852, Los Alamos, NM: Los Alamos National Lab.

[Huang & Vafai, 1994] Huang, P. & Vafai, K. (1994). Internal heat transfer augmentation in a channel using an alternate set of porous cavity-block obstacles. *Numer. Heat Transfer A*, 25, 519–540.

[Hughes et al., 1986] Hughes, T., Franca, L., & Balestra, M. (1986). A new finite element formulation for computational fluid dynamics: III. The general streamline operator for multidimensional advective-diffusion systems. *Comp. Meth. Appl. Mech. Engrg.*, 58, 305–328.

[Il'in, 1969] Il'in, A. (1969). Differencing scheme for a differential equation with small parameter affecting the highest derivative. *Math. Notes. Acad. Sci. USSR*, 6, 596–602.

[Jäger, 1982] Jäger, W. (1982). *Oszillatorische und turbulente Konvektion.* Dissertation, Universität Karlsruhe.

[Jakirlic & Hanjalic, 1994] Jakirlic, S. & Hanjalic, K. (1994). On the peformance of the second moment high and low Reynolds number closures in reattaching flows. In *Internat. Symposium on Turbulence, Heat and Mass Transfer*, Lisboa.

[Johnson, 1987] Johnson, C. (1987). *Numerical Solutions of Partial Differential Equations by the Finite Element Method.* Cambridge, UK: Cambridge University Press.

[Kaiktsis et al., 1991] Kaiktsis, L., Karniadakis, G., & Orszag, S. (1991). Onset of three-dimensionality, equilibra, and early transition in flow over a backward-facing step. *J. Fluid Mech.*, 231, 501–528.

[Kaufmann & Smarr, 1993] Kaufmann, W. & Smarr, L. (1993). *Supercomputing and the Transformation of Science.* New York: Scientific American Library. Distributed by W.H. Freeman.

[Kee et al., 1987] Kee, R., Rupley, F., & Miller, J. (1987). *The Chemkin thermodynamic data base.* Technical report SAND87-8215, Sandia National Laboratories, Albuquerque, NM.

[Keller & Keller, 1993] Keller, P. & Keller, M. (1993). *Visual Cues, Practical Data Visualization.* Hong Kong: IEEE Computer Society Press.

[Kessler, 1987] Kessler, R. (1987). Nonlinear transition in three-dimensional convection. *J. Fluid Mech.*, 174, 357–379.

[Khallouf et al., 1995] Khallouf, H., Gershuni, G., & Mojtabi, A. (1995). Numerical study of two-dimensional thermovibrational convection in rectangular cavities. *Numer. Heat Transfer A*, 27, 297–306.

[Kim et al., 1980] Kim, J., Kline, S., & Johnston, J. (1980). Investigations of a reattaching shear layer: Flows over a backward facing step. *ASME J. Fluids Engrg.*, 102, 302ff.

[Kim & Moin, 1985] Kim, J. & Moin, P. (1985). Application of a fractional-step method to incompressible Navier–Stokes equations. *J. Comput. Phys.*, 58, 308–323.

[Kim & Choudhury, 1993] Kim, S.-E. & Choudhury, D. (1993). A computational study of laminar flows: Benchmark calculations using FLUENT. *The CFD-Triathlon—Three Laminar Flow Simulations by Commercial CFD Codes*, 160, 23–34.

[Kimura & Bejan, 1983] Kimura, S. & Bejan, A. (1983). The heatline visualization of convective heat transfer. *Trans. ASME: J. Heat Transfer*, 105, 916–919.

[Kistler, 1983] Kistler, S. (1983). *The Fluid Mechanics of Curtain Coating and Related Viscous Free Surface Flows with Contact Lines.* Ph. D. Thesis, University of Minnesota.

[Knapek, 1994] Knapek, S. (1994). *Modellierung von porösen Medien mittels der Navier-Stokes-Gleichungen und statistisch generierter Hindernisse.* Fortgeschrittenenpraktikum, Institut für Informatik, TU München.

[Kolmogorov, 1942] Kolmogorov, A. (1942). The equations of turbulent motion in an incompressible fluid. *Izv. Sci. USSR Phys.*, 6, 56–58.

[Krause, 1994] Krause, E. (1994). High performance computing in fluid mechanics. In M. Griebel & C. Zenger (Eds.), *Numerical Simulation in Science and Engineering.* Braunschweig: Vieweg, 99–114.

[Ladyshenskaja, 1969] Ladyshenskaja, O. (1969). *The Mathematical Theory of Viscous Incompressible Flow.* New York: Gordon and Breach. Translation from Russian.

[Lafaurie et al., 1994] Lafaurie, B., Nardone, C., Scardovelli, R., Zaleski, S., & Zanetti, G. (1994). Modelling merging and fragmentation in multiphase flows with SURFER. *J. Comput. Phys.*, 113, 134–147.

[Lakhal et al., 1995] Lakhal, E., Hasnaoui, M., Vasseur, R., & Bilgen, E. (1995). Natural convection in a square enclosure heated periodically from part of the bottom wall. *Numer. Heat Transfer A*, 27, 319–334.

[Lam & Bremhorst, 1981] Lam, C. & Bremhorst, A. (1981). Modified form of the k-ε-model for predicting wall turbulence. *J. Fluids Engrg.*, 103, 456–460.

[Landau, 1959] Landau, L. (1959). *Course of Theoretical Physics/6. Fluid Mechanics.* Oxford: Pergamon Press.

[Launder, 1990] Launder, B. (1990). Phenomenological modelling: Present and future. In J. Lumley (Ed.), *Whither Turbulence: Turbulence at the Crossroads*. Berlin: Springer-Verlag.

[Launder et al., 1975] Launder, B., Reece, G., & Rodi, W. (1975). Progress in the development of a Reynolds stress turbulence closure. *J. Fluid Mech.*, 68.

[Launder & Spalding, 1972] Launder, B. & Spalding, D. (1972). *Mathematical Models of Turbulence*. New York: Academic Press.

[Leister, 1994] Leister, H.-J. (1994). *Numerische Simulation dreidimensionaler, zeitabhängiger Strömungen unter dem Einfluß von Auftriebs- und Trägheitskräften.* Dissertation, Universität Erlangen-Nürnberg.

[Leonard, 1979] Leonard, B. (1979). A stable and accurate convective modelling procedure based on quadratic upstream interpolation. *Comput. Meth. Appl. Mech. Engrg.*, 19, 59–98.

[Leschziner, 1989] Leschziner, M. (1989). Second moment closure for complex flows. In *Proc. Internat. Forum on Math. Mod. of Processes in Energy Systems*, Sarajevo.

[Macagno, 1965] Macagno, E. (1965). Some new aspects of similarity in hydraulics. *La Houille Blanche*, 20, 751–759.

[Mallinson & de Vahl Davis, 1973] Mallinson, G. & de Vahl Davis, G. (1973). The method of the false transient for the solution of coupled elliptic equations. *J. Comput. Phys.*, 12, 435–461.

[Mallinson & de Vahl Davis, 1977] Mallinson, G. & de Vahl Davis, G. (1977). Three-dimensional natural convection in a box: A numerical study. *J. Fluid Mech.*, 83, 1–31.

[McCormick, 1989] McCormick, S. (1989). *Multilevel Adaptive Methods for Partial Differential Equations*. Philadelphia: SIAM.

[Meier, 1995] Meier, F. (1995). *Numerische Simulation der Fluid-Struktur-Wechselwirkung am Beispiel einer ventilgesteuerten Mikropumpe.* Diplomarbeit, Institut für Informatik, TU München.

[Meneveau, 1991] Meneveau, C. (1991). Analysis of turbulence in the orthonormal wavelet representation. *J. Fluid Mech.*, 232, 469–520.

[Miyata, 1986] Miyata, H. (1986). Finite-difference simulation of breaking waves. *J. Comput. Phys.*, 65, 179–214.

[Miyata & Masuko, 1985] Miyata, H. & Masuko, A. (1985). Finite-difference simulation of nonlinear waves generated by ships of arbitrary three-dimensional configuration. *J. Comput. Phys.*, 60, 391–436.

[Moffatt, 1964] Moffatt, K. (1964). Viscous and resistive eddies near a sharp corner. *J. Fluid Mech.*, 18, 1–18.

[Mohammadi, 1992] Mohammadi, B. (1992). Complex turbulent flows computation with a two-layer approach. *Internat. J. Numer. Methods Fluids*, 15, 747–771.

[Mohammadi & Pironneau, 1993] Mohammadi, B. & Pironneau, O. (1993). *Analysis of the K-Epsilon Turbulence Model.* Chichester, UK: Wiley.

[Nag et al., 1994] Nag, A., Sarkar, A., & Sastri, V. (1994). Effect of thick horizontal partial partition attached to one of the active walls of a differentially heated square cavity. *Numer. Heat Transfer A*, 25, 611–626.

[Nakayama, 1988] Nakayama, Y. (1988). *Visualized Flow.* Oxford: Pergamon Press.

[Neumann, 1990] Neumann, G. (1990). Three-dimensional numerical simulation of buoyancy-driven convection in vertical cylinders heated from below. *J. Fluid Mech.*, 214, 559–578.

[Neunhoeffer, 1997] Neunhoeffer, T. (1997). *Numerische Simulation von Erstarrungsprozessen unterkühlter Flüssigkeiten unter Berücksichtigung von Dichteunterschieden,* Dissertation, Institut für Informatik, TU München.

[Nichols & Hirt, 1971] Nichols, B. & Hirt, C. (1971). Improved free surface boundary-conditions for numerical incompressible-flow calculations. *J. Comput. Phys.*, 8, 434–448.

[Oberbeck, 1879] Oberbeck, A. (1879). Über die Wärmeleitung der Flüssigkeiten bei Berücksichtigung der Strömungen infolge von Temperaturdifferenzen. *Ann. Phys. Chem.*, 7, 271–292.

[Oran & Boris, 1987] Oran, E. & Boris, J. (1987). *Numerical Simulation of Reactive Flow.* New York: Elsevier.

[Ozal & Hara, 1995] Ozal, H. & Hara, T. (1995). Numerical analysis for oscillatory natural convection of low Prandtl number fluid heated from below. *Numer. Heat Transfer A*, 27, 307–318.

[Patankar, 1980] Patankar, S. (1980). *Numerical Heat Transfer and Fluid Flow.* New York: McGraw-Hill.

[Patankar & Spalding, 1972] Patankar, S. & Spalding, D. (1972). A calculation procedure for heat, mass- and momentum transfer in three-dimensional parabolic flows. *J. Heat Mass Transfer*, 15, 1787–1806.

[Patel et al., 1989] Patel, V., Rodi, W., & Scheuerer, G. (1989). Turbulence models for near-wall and low Reynolds number flows: A review. *AIAA J.*, 23, 1308–1319.

[Pavlidis, 1982] Pavlidis, T. (1982). *Graphics and Image Processing.* Berlin, Heidelberg: Springer-Verlag.

[Peaceman & Rachford, 1955] Peaceman, D. & Rachford, H. (1955). The numerical solution of parabolic and elliptic differential equations. *J. Soc. Indust. Appl. Math.*, 3, 28–41.

[Pelz et al., 1993] Pelz, R., Ecer, A., & Häuser, J., Eds. (1993). *Parallel Computational Fluid Dynamics* '92. Amsterdam: North Holland.

[Perić et al., 1988] Perić, M., Kessler, R., & Scheuerer, G. (1988). Comparison of finite-volume numerical methods with staggered and colocated grids. *Comput. & Fluids*, 16, 389–403.

[Perry et al., 1982] Perry, A., Chong, M., & Lim, T. (1982). The vortex-shedding process behind two-dimensional bluff bodies. *J. Fluid Mech.*, 116, 77–90.

[Peyret & Taylor, 1983] Peyret, R. & Taylor, T. (1983). *Computational Methods for Fluid Flow.* Springer Ser. Comput. Phys. Berlin: Springer-Verlag.

[Pinelli & Vacca, 1994] Pinelli, A. & Vacca, A. (1994). Chebyshev collocation method and multidomain decomposition for the incompressible Navier-Stokes-equations. *Internat. J. Numer. Methods Fluids*, 18, 781–799.

[Prandtl, 1936a] Prandtl, L. (1936a). *Entstehung von Wirbeln bei Wasserströmungen, Teil* 1: *Entstehung von Wirbeln.* Göttingen: Institut für den Wissenschaftlichen Film.

[Prandtl, 1936b] Prandtl, L. (1936b). *Entstehung von Wirbeln bei Wasserströmungen, Teil* 2: *Anwendungen auf die Strömungen.* Göttingen: Institut für den Wissenschaftlichen Film.

[Preparata & Shamos, 1985] Preparata, F. & Shamos, M. (1985). *Computational Geometry.* Berlin, Heidelberg: Springer-Verlag.

[Press et al., 1990] Press, W., Flannery, B., Teukolsky, S., & Vetterling, W. (1990). *Numerical Recipes in C.* Cambridge, UK: Cambridge University Press.

[Quateroni, 1991] Quateroni, A. (1991). Domain decomposition and parallel processing for the numerical solution of partial differential equations. *Surveys Math. Industry*, 1, 75–118.

[Rannacher, 1992] Rannacher, R. (1992). *On the Numerical Solution of the Incompressible Navier-Stokes-Equations.* Preprint 92–16, IWR Heidelberg.

[Reichert & Wittum, 1994] Reichert, H. & Wittum, G. (1994). *Robust Multigrid Methods for the Incompressible Navier-Stokes-Equations.* Bericht nr 94/2, Inst. für Computeranwendungen der Universität Stuttgart.

[Richardson & Cornish, 1977] Richardson, S. & Cornish, A. (1977). Solution of three-dimensional incompressible flow problems. *J. Fluid Mech.*, 82, 309–340.

[Roache, 1976] Roache, P. (1976). *Computational Fluid Dynamics.* Albuquerque: Hermosa.

[Rodi, 1994] Rodi, W. (1994). Current trends in turbulence modelling. In L. Fezoni, J. Periaux, & J. Hunt (Eds.), *Computational Aeronautical Fluid Dynamics.* Oxford: Clarendon Press, 225–245.

[Rosenfeld et al., 1991] Rosenfeld, M., Kwak, D., & Vinokur, M. (1991). A fractional step method for the unsteady incompressible Navier-Stokes-equations in generalized coordinate systems. *J. Comput. Phys.*, 94, 102–137.

[Russell, 1992] Russell, T., Ed. (1992). *Computational Methods in Water Resources* IX, *Vol.* 2: *Mathematical Modeling in Water Resources.* Southampton: Computational Mechanics.

[Sakamoto & Matsuo, 1980] Sakamoto, Y. & Matsuo, Y. (1980). Numerical prediction of three-dimensional flow in a ventilated room using turbulence models. *Appl. Math. Modelling*, 4, 67–72.

[Schäfer, 1994] Schäfer, M. (1994). Efficient methods and parallel computing in numerical fluid dynamics. In M. Griebel & C. Zenger (Eds.), *Numerical Simulation in Science and Engineering.* Braunschweig: Vieweg, 173–188.

[Schlichting, 1982] Schlichting, H. (1982). *Boundary-Layer Theory.* New York: McGraw-Hill.

[Schmitt, 1988] Schmitt, L. (1988). *Grobstruktursimulation turbulenter Grenzschicht-, Kanal- und Stufenströmungen.* Dissertation, Lehrstuhl für Strömungsmechanik, TU München.

[Schumann, 1977] Schumann, V. (1977). Realizabilty of Reynolds stress turbulence models. *Phys. Fluids*, 20, 721–725.

[Schwarz, 1890] Schwarz, H. (1890). *Gesammelte mathematische Abhandlungen, Band 2.* Berlin: Springer-Verlag.

[Simon, 1992] Simon, H., Ed. (1992). *Parallel Computational Fluid Dynamics, Implementation and Results.* Cambridge, MA: MIT Press.

[Smagorinsky, 1963] Smagorinsky, J. (1963). General circulation model of the atmosphere. *Mon. Weather Rev.*, 91, 99–164.

[Smith et al., 1995] Smith, B., Bjørstad, P., & Gropp, W. (1995). *Domain Decomposition: Parallel Multilevel Algorithms for Elliptic Partial Differential Equations.* Cambridge, UK: Cambridge University Press.

[Sod, 1985] Sod, G. (1985). *Numerical Methods for Fluid Dynamics.* Cambridge, UK: Cambridge University Press.

[Sohn, 1988] Sohn, J. (1988). Evaluation of FIDAP on some classical laminar and turbulent benchmarks. *J. Numer. Methods Fluids*, 8, 1469–1490.

[Stevens, 1990] Stevens, W. (1990). *Unix Network Programming.* Englewood Cliffs, NJ: Prentice Hall.

[Stoer & Bulirsch, 1980] Stoer, J. & Bulirsch, R. (1980). *Introduction to Numerical Analysis.* Berlin: Springer-Verlag.

[Strang & Fix, 1973] Strang, G. & Fix, G. (1973). *An Analysis of the Finite Element Method.* Englewood Cliffs, NJ: Prentice Hall.

[Strumpen, 1995] Strumpen, V. (1995). Coupling hundreds of workstations for parallel molecular sequence analysis. *Software—Practice and Experience*, 25, 291–304.

[Temam, 1969] Temam, R. (1969). Sur l'approximation de la solution des équations de Navier–Stokes par la méthode des pas fractionnaires. *Arch. Rational Mech. Anal.*, 32, 135–153.

[Tennekes & Lumley, 1972] Tennekes, H. & Lumley, J. (1972). *A First Course in Turbulence.* Cambridge, MA: MIT Press.

[Thom, 1933] Thom, A. (1933). The flow past circular cylinders at low speeds. *Proc. Roy. Soc. London Ser. A*, 141, 651–666.

[Tome & McKee, 1994] Tome, M. & McKee, S. (1994). GENSMAC: A computational marker and cell method for free surface flows in general domains. *J. Comput. Phys.*, 110, 171–186.

[Trefil, 1986] Trefil, J. (1986). *Meditations at* 10000 *Feet: A Scientist in the Mountains.* New York: C. Scribner's Sons.

[Tritton, 1959] Tritton, D. (1959). Experiments on the flow past a circular cylinder at low Reynolds numbers. *J. Fluid Mech.*, 6, 547–567.

[Turek, 1992] Turek, S. (1992). *Tools for Simulating Nonstationary Incompressible Flow via Discretely Divergence-Free Finite Element Methods.* SFB-Bericht 679, Universität Heidelberg.

[van Dyke, 1982] van Dyke, M. (1982). *An Album of Fluid Motion.* Stanford, CA: The Parabolic Press.

[Varga, 1962] Varga, R. (1962). *Matrix Iterative Analysis.* Englewood Cliffs, NJ: Prentice Hall.

[Viecelli, 1969] Viecelli, J. (1969). A method for including arbitrary external boundaries in the MAC incompressible fluid computing technique. *J. Comput. Phys.*, 4, 543–551.

[Viecelli, 1971] Viecelli, J. (1971). A computing method for incompressible flows bounded by moving walls. *J. Comput. Phys.*, 8, 119–143.

[Viollet, 1981] Viollet, P. (1981). On the modeling of turbulent heat and mass transfers for computations of buoyancy affected flows. *Proc. Internat. Conf. Num. Meth. for Laminar and Turbulent Flows.*

[Wesseling, 1992] Wesseling, P. (1992). *An Introduction to Multigrid Methods.* Chichester, UK: Wiley.

[Wong & Reizes, 1984] Wong, A. & Reizes, J. (1984). An effective vorticity-vector potential formulation for the numerical solution of three-dimensional duct flow problems. *J. Comput. Phys.*, 55, 98–114.

[Wung & Tseng, 1992] Wung, T.-S. & Tseng, F.-G. (1992). A color-coded particle tracking velocimeter with application to natural convection. *Experiments in Fluids*, 13, 217–223.

[Yoo et al., 1994] Yoo, J.-S., Choi, J. & Kim, M.-U. (1994). Multicellular natural convection of a low Prandtl number fluid between concentric cylinders. *Numer. Heat Transfer A*, 25, 103–115.

Index